Keynotes in Organic Chemistry

Keynotes in Organic Chemistry

Second Edition

ANDREW F. PARSONS
Department of Chemistry, University of York, UK

This edition first published 2014

© 2014 John Wiley & Sons, Ltd

Registered office

John Wiley & Sons Ltd, The Atrium, Southern Gate, Chichester, West Sussex, PO19 8SQ, United Kingdom

For details of our global editorial offices, for customer services and for information about how to apply for permission to reuse the copyright material in this book please see our website at www.wiley.com.

Library of Congress Cataloging-in-Publication Data

Parsons, A. F.
    Keynotes in organic chemistry / Andrew Parsons. – Second edition.
        pages cm.
    Includes bibliographical references and index.
    ISBN 978-1-119-99915-7   (hardback) – ISBN 978-1-119-99914-0 (paperback)   1.
Chemistry, Organic–Outlines, syllabi, etc.   I. Title.
    QD256.5.P35 2014
    547–dc23   2013024694

A catalogue record for this book is available from the British Library.

HB ISBN: 9781119999157
PB ISBN: 9781119999140

Set in 10/12pt Times by Thomson Digital, Noida, India.

Printed and bound in Singapore by Markono Print Media Pte Ltd

1   2014

Contents

Preface

With the advent of modularisation and an ever-increasing number of examinations, there is a growing need for concise revision notes that encapsulate the key points of a subject in a meaningful fashion. This keynote revision guide provides concise organic chemistry notes for first year students studying chemistry and related courses (including biochemistry) in the UK. The text will also be appropriate for students on similar courses in other countries.

An emphasis is placed on presenting the material pictorially (pictures speak louder than words); hence, there are relatively few paragraphs of text but numerous diagrams. These are annotated with key phrases that summarise important concepts/key information and bullet points are included to concisely highlight key principles and definitions.

The material is organised to provide a structured programme of revision. Fundamental concepts, such as structure and bonding, functional group identification and stereochemistry are introduced in the first three chapters. An important chapter on reactivity and mechanism is included to provide a short overview of the basic principles of organic reactions. The aim here is to provide the reader with a summary of the 'key tools' which are necessary for understanding the following chapters and an important emphasis is placed on organisation of material based on reaction mechanism. Thus, an overview of general reaction pathways/mechanisms (such as substitution and addition) is included and these mechanisms are revisited in more detail in the following chapters. Chapters 5–10 are treated essentially as 'case studies', reviewing the chemistry of the most important functional groups. Halogenoalkanes are discussed first and as these compounds undergo elimination reactions this is followed by the (electrophilic addition) reactions of alkenes and alkynes. This leads on to the contrasting (electrophilic substitution) reactivity of benzene and derivatives in Chapter 7, while the rich chemistry of carbonyl compounds is divided into Chapters 8 and 9. This division is made on the basis of the different reactivity (addition versus substitution) of aldehydes/ketones and carboxylic acid derivatives to nucleophiles. A chapter is included to revise the importance of spectroscopy in structure elucidation and, finally, the structure and reactivity of a number of important natural products and synthetic polymers is highlighted in Chapter 11. Worked examples and questions are included at the end of each chapter to test the reader's understanding, and outline answers are provided for all of the questions. Tables of useful physical data, reaction summaries and a glossary are included in appendices at the back of the book.

New to this edition

A number of additions have been made to this edition to reflect the feedback from students and lecturers:

- A second colour is used to clarify some of the diagrams, particularly the mechanistic aspects
- Reference notes are added in the margin to help the reader find information and to emphasise links between different topics
- Diagrams are included in the introductory key point sections for each chapter
- Additional end-of-chapter problems (with outline answers) are included
- A worked example is included at the end of each chapter
- The information in the appendices has been expanded, including reaction summaries and a glossary

Acknowledgements

There are numerous people I would like to thank for their help with this project. This includes many students and colleagues at York. Their constructive comments were invaluable. I would also like to thank my family for their support and patience throughout this project. Finally, I would like to thank Paul Deards and Sarah Tilley from Wiley, for all their help in progressing the second edition.

Dr Andrew F. Parsons
2013

1

Structure and bonding

Key point. Organic chemistry is the study of carbon compounds. *Ionic* bonds involve elements gaining or losing electrons but the carbon atom is able to form four *covalent* bonds by sharing the four electrons in its outer shell. Single (C–C), double (C=C) or triple bonds (C≡C) to carbon are possible. When carbon is bonded to a different element, the electrons are not shared equally, as *electronegative* atoms (or groups) attract the electron density whereas *electropositive* atoms (or groups) repel the electron density. An understanding of the electron-withdrawing or -donating ability of atoms, or a group of atoms, can be used to predict whether an organic compound is a good *acid* or *base*.

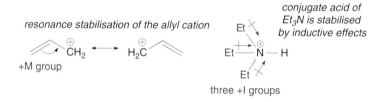

1.1 Ionic versus covalent bonds

- *Ionic bonds* are formed between molecules with opposite charges. The negatively charged anion will electrostatically attract the positively charged cation. This is present in (inorganic) salts.

$$\text{Cation}^{\oplus}\!\!\!\ldots\ldots\ldots^{\ominus}\text{Anion} \qquad \text{e.g.} \qquad \text{Na}^{\oplus}\!\!\!\ldots\ldots\ldots^{\ominus}\text{Cl}$$

- *Covalent bonds* are formed when a pair of electrons is shared between two atoms. A single line represents the two-electron bond.

$$\text{Atom}\!\!-\!\!\text{Atom} \qquad \text{e.g.} \quad \text{Cl}-\text{Cl} \; \equiv \; \text{Cl}\;\text{Cl}$$

Keynotes in Organic Chemistry, Second Edition. Andrew F. Parsons.
© 2014 John Wiley & Sons, Ltd. Published 2014 by John Wiley & Sons, Ltd.

- *Coordinate (or dative) bonds* are formed when a pair of electrons is shared between two atoms. *One* atom donates both electrons and a single line or an arrow represents the two-electron bond.

The cyclic ether is tetrahydrofuran (THF) and BH_3 is called borane (Section 6.2.2.5)

electron acceptor

O ——→ BH₃   or   O — BH₃

electron donor

- *Hydrogen bonds* are formed when the partially positive ($\delta+$) hydrogen of one molecule interacts with the partially negative ($\delta-$) heteroatom (e.g. oxygen or nitrogen) of another molecule.

Intramolecular hydrogen bonding in carbonyl compounds is discussed in Section 8.4.1

$$\overset{\delta+}{\text{Molecule–H}}\ \cdots\ \overset{\delta-}{\text{Heteroatom–Molecule}} \quad \text{e.g.} \quad \text{HO}\ \overset{\delta+}{\text{—H}}\ \cdots\ \overset{\delta-}{\text{OH}_2}$$

1.2 The octet rule

To form organic compounds, the carbon atom shares electrons to give a stable 'full shell' electron configuration of eight valence electrons.

Methane is the smallest alkane – alkanes are a family of compounds that contain only C and H atoms linked by single bonds (Section 2.4)

Methane (CH₄)

Lewis structure

C is in group 14 and so has 4 valence electrons
H is in group 1 and so has 1 valence electron

Full structural formula (or Kekulé structure)
A line = 2 electrons

Drawing organic compounds using full structural formulae and other conventions is discussed in Section 2.5

A single bond contains two electrons, a double bond contains four electrons and a triple bond contains six electrons. A lone (or non-bonding) pair of electrons is represented by two dots (· ·).

Carbon dioxide (CO₂)   *Hydrogen cyanide (HCN)*

1.3 Formal charge

Formal positive or negative charges are assigned to atoms, which have an apparent 'abnormal' number of bonds.

Atom(s)	C	N, P	O, S	F, Cl, Br, I
Group number	14	15	16	17
Normal number of 2 electron bonds	4	3	2	1

Formal charge = group number in periodic table − number of bonds to atom − number of unshared electrons − 10

Example: Nitric acid (HNO₃)

Nitrogen with 4 covalent bonds has a formal charge of +1

Formal charge: 15 − 4 − 0 − 10 = +1

Nitric acid is used in synthesis to nitrate aromatic compounds such as benzene (Section 7.2.2)

The nitrogen atom donates a pair of electrons to make this bond

Carbon forms four covalent bonds. When only three covalent bonds are present, the carbon atom can have either a formal negative charge or a formal positive charge.

The stability of carbocations and carbanions is discussed in Section 4.3

- *Carbanions*–three covalent bonds to carbon and a formal negative charge.

Formal charge on C:
14 − 3 − 2 − 10 = −1

8 outer electrons: 3 two-electron bonds and 2 non-bonding electrons

Carbanions are formed on deprotonation of organic compounds. Deprotonation of a carbonyl compound, at the α-position, forms a carbanion called an enolate ion (Section 8.4.3)

The negative charge is used to show the 2 non-bonding electrons

- *Carbocations*–three covalent bonds to carbon and a formal positive charge.

Formal charge on C:
14 − 3 − 0 − 10 = +1

6 outer electrons: 3 two-electron bonds

Carbocations are intermediates in a number of reactions, including S_N1 reactions (Section 5.3.1.2)

The positive charge is used to show the absence of 2 electrons

1.4 Sigma (σ−) and pi (π−) bonds

The electrons shared in a covalent bond result from overlap of atomic orbitals to give a new molecular orbital. Electrons in 1s and 2s orbitals combine to give sigma (σ−) bonds.

When two 1s orbitals combine *in-phase*, this produces a *bonding molecular orbital*.

Molecular orbitals and chemical reactions are discussed in Section 4.10

s-orbital   +   s-orbital   ⟶   bonding molecular orbital

When two 1s orbitals combine *out-of-phase*, this produces an *antibonding molecular orbital*.

s-orbital   +   s-orbital   ⟶   antibonding molecular orbital

Electrons in p orbitals can combine to give sigma (σ) or pi (π) bonds.

- *Sigma ($\sigma-$) bonds* are strong bonds formed by head-on overlap of two atomic orbitals.

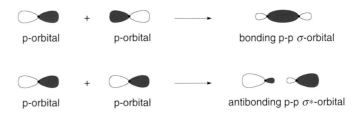

p-orbital   +   p-orbital   ⟶   bonding p-p σ-orbital

p-orbital   +   p-orbital   ⟶   antibonding p-p $\sigma*$-orbital

Alkenes have a C=C bond containing one strong σ-bond and one weaker π-bond (Section 6.1)

- *Pi ($\pi-$) bonds* are weaker bonds formed by side-on overlap of two p-orbitals.

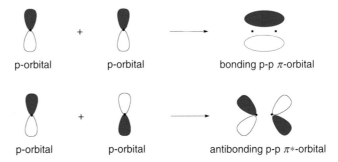

p-orbital   +   p-orbital   ⟶   bonding p-p π-orbital

All carbonyl compounds have a C=O bond, which contains one strong σ-bond and one weaker π-bond (Section 8.1)

p-orbital   +   p-orbital   ⟶   antibonding p-p $\pi*$-orbital

Only σ- or π-bonds are present in organic compounds. All single bonds are σ-bonds while all multiple (double or triple) bonds are composed of one σ-bond and one or two π-bonds.

1.5 Hybridisation

Hund's rule states that when filling up a set of orbitals of the same energy, electrons are added with parallel spins to different orbitals rather than pairing two electrons in one orbital

- The ground-state electronic configuration of carbon is $1s^2 2s^2 2p_x{}^1 2p_y{}^1$.
- The six electrons fill up lower energy orbitals before entering higher energy orbitals (Aufbau principle).
- Each orbital is allowed a maximum of two electrons (Pauli exclusion principle).
- The two 2p electrons occupy separate orbitals before pairing up (Hund's rule).

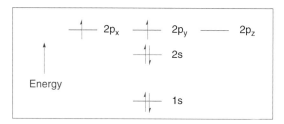

The carbon atom can mix the 2s and 2p atomic orbitals to form four new hybrid orbitals in a process known as *hybridisation*.

- *sp^3 Hybridisation*. For four single σ-bonds – carbon is sp^3 hybridised (e.g. in methane, CH_4). The orbitals move as far apart as possible, and the lobes point to the corners of a tetrahedron (109.5° bond angle).

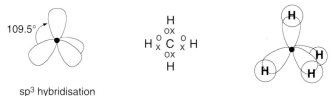

sp3 hybridisation

methane: 4 × C–H σ-bonds

- *sp^2 Hybridisation*. For three single σ-bonds and one π-bond – the π-bond requires one p-orbital, and hence the carbon is sp^2 hybridised (e.g. in ethene, $H_2C{=}CH_2$). The three sp^2-orbitals point to the corners of a triangle (120° bond angle), and the remaining p-orbital is perpendicular to the sp^2 plane.

Alkenes have a C=C bond containing one strong σ-bond and one weaker π-bond (Section 6.1)

All carbonyl compounds have a C=O bond, which contains one strong σ-bond and one weaker π-bond (Section 8.1)

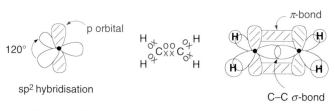

sp2 hybridisation

ethene: 4 × C–H σ-bonds, 1 × C–C σ-bond, 1 × C–C π-bond

- *sp Hybridisation*. For two single σ-bonds and two π-bonds – the two π-bonds require two p-orbitals, and hence the carbon is sp hybridised (e.g. in ethyne, HC≡CH). The two sp-orbitals point in the opposite directions (180° bond angle), and the two p-orbitals are perpendicular to the sp plane.

Alkynes have a C≡C bond containing one strong σ-bond and two weaker π-bonds (Section 6.1)

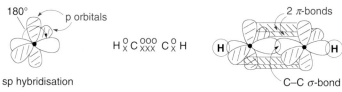

sp hybridisation

ethyne: 2 × C–H σ-bonds, 1 × C–C σ-bond, 2 × C–C π-bonds

- For a single C–C or C–O bond, the atoms are sp^3 hybridised and the carbon atom(s) is *tetrahedral*.
- For a double C=C or C=O bond, the atoms are sp^2 hybridised and the carbon atom(s) is *trigonal planar*.
- For a triple C≡C or C≡N bond, the atoms are sp hybridised and the carbon atom(s) is *linear*.

This compound contains four functional groups, including a phenol. Functional groups are introduced in Section 2.1

3 = sp3 2 = sp2 1 = sp

The shape of organic molecules is therefore determined by the hybridisation of the atoms.

Functional groups (Section 2.1) that contain π-bonds are generally more reactive as a π-bond is weaker than a σ-bond. The π-bond in an alkene or alkyne is around +250 kJ mol^{-1}, while the σ-bond is around +350 kJ mol^{-1}.

Bond	Mean bond enthalpies (kJ mol^{-1})	Mean bond lengths (pm)
C—C	+347	153
C=C	+612	134
C≡C	+838	120

A hydrogen atom attached to a C≡C bond is more acidic than a hydrogen atom attached to a C=C bond or a C–C bond; this is explained by the change in hybridisation of the carbon atom that is bonded to the hydrogen atom (Section 1.7.4)

The shorter the bond length, the stronger the bond. For C–H bonds, the greater the 's' character of the carbon orbitals, the shorter the bond length. This is because the electrons are held closer to the nucleus.

sp3   sp2   sp
H$_3$C—CH$_2$ꓔH > H$_2$C=CH—H > HC≡CꓔH
longest   shortest

Rotation about C–C bonds is discussed in Section 3.2

A single C–C σ-bond can undergo free rotation at room temperature, but a π-bond prevents free rotation around a C=C bond. For maximum orbital overlap in a π-bond, the two p-orbitals need to be parallel to one another. Any rotation around the C=C bond will break the π-bond.

1.6 Inductive effects, hyperconjugation and mesomeric effects

1.6.1 Inductive effects

In a covalent bond between two different atoms, the electrons in the σ-bond are not shared equally. The electrons are attracted towards the most electronegative atom. An arrow drawn above the line representing the covalent bond can show this. (Sometimes an arrow is drawn on the line.) Electrons are pulled in the direction of the arrow.

When the atom (X) is more electronegative than carbon	When the atom (Z) is less electronegative than carbon
electrons **attracted to X**	electrons **attracted to C**
$\overset{\delta+}{-}\text{C}\overset{\delta-}{-}\text{X}$	$\overset{\delta-}{-}\text{C}\overset{\delta+}{-}\text{Z}$
negative inductive effect. –I	positive inductive effect. +I

–I groups	**+I groups**
X = Br, Cl, NO$_2$, OH, OR, SH, SR, NH$_2$, NHR, NR$_2$, CN, CO$_2$H, CHO, C(O)R	Z = R (alkyl or aryl), metals (e.g. Li or Mg)
The more electronegative the atom (X), the stronger the –I effect	The more electropositive the atom (Z), the stronger the +I effect

Pauling electronegativity scale	The inductive effect of the atom rapidly diminishes as the chain length increases
K = 0.82 I = 2.66 C = 2.55 Br = 2.96 N = 3.04 Cl = 3.16 O = 3.44 F = 3.98	$\underset{}{\text{H}_3\text{C}}\overset{\delta\delta\delta+}{-}\text{CH}_2\overset{\delta\delta+}{-}\text{CH}_2\overset{\delta+}{-}\text{CH}_2\overset{\delta-}{-}\text{Cl}$
The higher the value the more electronegative the atom	experiences a negligible –I effect experiences a strong –I effect

The overall polarity of a molecule is determined by the individual bond polarities, formal charges and lone pair contributions and this can be measured by the dipole moment (μ). The larger the dipole moment (often measured in debyes, D), the more polar the compound.

1.6.2 Hyperconjugation

A σ-bond can stabilise a neighbouring carbocation (or positively charged carbon, e.g. R$_3$C$^+$) by donating electrons to the vacant p-orbital. The positive charge is delocalised or 'spread out' and this stabilising effect is called *resonance*.

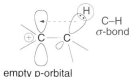

empty p-orbital

C–H σ-bond   The electrons in the C–H σ-bond spend some of the time in the empty p-orbital

1.6.3 Mesomeric effects

Whilst inductive effects pull electrons through the σ-bond framework, electrons can also move through the π-bond network. A π-bond can stabilise a negative charge, a

An inductive effect is the polarisation of electrons through σ-bonds

An alkyl group (R) is formed by removing a hydrogen atom from an alkane (Section 2.2).

An aryl group (Ar) is benzene (typically called phenyl, Ph) or a substituted benzene group (Section 2.2)

Hyperconjugation is the donation of electrons from nearby C–H or C–C σ-bonds

The stability of carbocations is discussed in Section 4.3.1

Resonance forms (sometimes called canonical forms) show all possible distributions of electrons in a molecule or an ion

positive charge, a lone pair of electrons or an adjacent bond by *resonance* (i.e. delocalisation or 'spreading out' of the electrons). Curly arrows (Section 4.1) are used to represent the movement of π- or non-bonding electrons to give different resonance forms. It is only the electrons, not the nuclei, that move in the resonance forms, and a double-headed arrow is used to show their relationship.

1.6.3.1 Positive mesomeric effect

- When a π-system donates electrons, the π-system has a positive mesomeric effect, $+M$.

This carbocation is called an allylic cation (see Section 5.3.1.2)

donates electrons:
+M group

- When a lone pair of electrons is donated, the group donating the electrons has a positive mesomeric effect, $+M$.

The OR group is called an alkoxy group (see Section 2.4)

donates electrons:
+M group

1.6.3.2 Negative mesomeric effect

- When a π-system accepts electrons, the π-system has a negative mesomeric effect, $-M$.

This anion, formed by deprotonating an aldehyde at the α-position, is called an enolate ion (Section 8.4.3)

accepts electrons:
–M groups

Functional groups are discussed in Section 2.1

The actual structures of the cations or anions lie somewhere between the two resonance forms. All resonance forms must have the same overall charge and obey the same rules of valency.

> **–M groups** generally contain an electronegative atom(s) and/or a π-bond(s): CHO, C(O)R, CO_2H, CO_2Me, NO_2, CN, aromatics, alkenes
>
> **+M groups** generally contain a lone pair of electrons or a π-bond(s): $\ddot{C}l$, $\ddot{B}r$, $\ddot{O}H$, $\ddot{O}R$, $\ddot{S}H$, $\ddot{S}R$, $\dot{N}H_2$, $\dot{N}HR$, $\dot{N}R_2$, aromatics, alkenes
>
> Aromatic (or aryl) groups and alkenes can be *both* +M and –M.

In neutral compounds, there will always be a $+M$ *and* $-M$ group(s): one group donates $(+M)$ the electrons, the other group(s) accepts the electrons $(-M)$.

An amide, such as $RCONH_2$, also contains both a $+M$ group (NH_2) and a $-M$ group (C=O). See Sections 1.7.2 and 9.3.1

$$R\ddot{O}-CH=CHR \longleftrightarrow R\overset{\oplus}{O}=CH-\overset{\ominus}{C}HR$$

+M group −M group

All resonance forms are *not* of the same energy. Generally, the most stable resonance forms have the greatest number of covalent bonds, atoms with a complete valence shell of electrons, and/or an aromatic ring. In phenol (PhOH), for example, the resonance form with the intact aromatic benzene ring is expected to predominate.

Benzene and other aromatic compounds, including phenol, are discussed in Chapter 7

aromatic
ring is intact

As a rule of thumb, the more resonance structures an anion, cation or neutral π-system can have, the more stable it is.

1.6.3.3 Inductive versus mesomeric effects

Mesomeric effects are generally stronger than inductive effects. A $+M$ group is likely to stabilise a cation more effectively than a $+I$ group.

Mesomeric effects can be effective over much longer distances than inductive effects provided that *conjugation* is present (i.e. alternating single and double bonds). Whereas inductive effects are determined by distance, mesomeric effects are determined by the relative positions of $+M$ and $-M$ groups in a molecule (Section 1.7).

Conjugated enones, containing a C=C–C=O group, are discussed in Section 8.5.1

1.7 Acidity and basicity

1.7.1 Acids

An acid is a substance that donates a proton (Brønsted-Lowry). Acidic compounds have low pK_a values and are good proton donors as the anions (or conjugate bases), formed on deprotonation, are relatively stable.

Equilibria and equilibrium constants are discussed in Section 4.9.1.1

In water:

$$HA + H_2\ddot{O} \underset{}{\overset{K_a}{\rightleftharpoons}} H_3\overset{\oplus}{O} + \overset{\ominus}{A}$$

Acid   Base   Conjugate   Conjugate
                         acid   base

acidity constant

The more stable the conjugate base the stronger the acid

$$K_a \approx \frac{[H_3\overset{\oplus}{O}][\overset{\ominus}{A}]}{[HA]}$$

As H_2O is in excess

$$pK_a = -\log_{10}K_a$$

The higher the value of K_a, the lower the pK_a value and the more acidic is HA

The pK_a value equals the pH of the acid when it is half ionised. At pH's above the pK_a the acid (HA) exists predominantly as the conjugate base (A$^-$) in water. At pH's below the pK_a it exists predominantly as HA.

> pH = 0, strongly acidic
>
> pH = 7, neutral
>
> pH = 14, strongly basic

The influence of solvent polarity on substitution and elimination reactions is discussed in Sections 5.3.1.3 and 5.3.2.3

The pK_a values are influenced by the solvent. Polar solvents will stabilise cations and/or anions by *solvation* in which the charge is delocalised over the solvent (e.g. by hydrogen-bonding in water).

The more electronegative the atom bearing the negative charge, the more stable the conjugate base (which is negatively charged).

pK_a	3		16		33		48	
most acidic	HF	>	H$_2$O	>	NH$_3$	>	CH$_4$	*least acidic*

decreasing electronegativity on going from F to C

Therefore, F$^-$ is more stable than H$_3$C$^-$.

Inductive effects are introduced in Section 1.6.1

The conjugate base can also be stabilised by $-$I and $-$M groups which can delocalise the negative charge. (The more 'spread out' the negative charge, the more stable it is).

Mesomeric effects are introduced in Section 1.6.3

> $-$I and $-$M groups therefore *lower* the pK_a while +I and +M groups *raise* the pK_a

1.7.1.1   Inductive effects and carboxylic acids

The carboxylate ion (RCO$_2^-$) is formed on deprotonation of a carboxylic acid (RCO$_2$H). The anion is stabilised by resonance (i.e. the charge is spread over both oxygen atoms) but can also be stabilised by the R group if this has a $-$I effect.

The reactions of carboxylic acids are discussed in Chapter 9

The greater the $-$I effect, the more stable the carboxylate ion (e.g. FCH$_2$CO$_2^-$ is more stable than BrCH$_2$CO$_2^-$) and the more acidic the carboxylic acid (e.g. FCH$_2$CO$_2$H is more acidic than BrCH$_2$CO$_2$H).

$$F-CH_2-CO_2H \qquad Br-CH_2-CO_2H \qquad H_3C-CO_2H$$

pK$_a$　　　2.7　　　　　　　　2.9　　　　　　　　4.8

Most acidic as F is more　　　　　　　　　　**Least** acidic as the CH$_3$
electronegative than Br and　　　　　　　　　　group is a +I group
has a greater −I effect

1.7.1.2　Inductive and mesomeric effects and phenols

Mesomeric effects can also stabilise positive and negative charges.

> The *negative* charge needs to be on an adjacent carbon atom
> for a **−M** group to stabilise it
>
> The *positive* charge needs to be on an adjacent carbon atom
> for a **+M** group to stabilise it

On deprotonation of phenol (PhOH) the phenoxide ion (PhO$^-$) is formed. This anion is stabilised by the delocalisation of the negative charge on to the 2-, 4- and 6-positions of the benzene ring.

- If −M groups are introduced at the 2-, 4- and/or 6-positions, the anion can be further stabilised by delocalisation through the π-system as the negative charge can be spread onto the −M group. We can use double-headed curly arrows to show this process.

  Double-headed curly arrows are introduced in Section 4.1

- If −M groups are introduced at the 3- and/or 5-positions, the anion cannot be stabilised by delocalisation, as the negative charge cannot be spread onto the −M group. There is no way of using curly arrows to delocalise the charge on to the −M group.
- If −I groups are introduced on the benzene ring, the effect will depend on their distance from the negative charge. The closer the −I group is to the negative charge, the greater the stabilising effect will be. The order of −I stabilisation is therefore 2-position > 3-position > 4-position.
- The −M effects are much stronger than −I effects (Section 1.6.3).

Examples

> The NO$_2$ group is strongly electron-withdrawing; −I and −M

Naming substituted benzenes is
discussed in Section 2.4

pK_a 9.9	8.4	7.2	4.0
Least acidic as no $-I$ or $-M$ groups on the ring	The NO_2 can only stabilise the anion inductively	The NO_2 can stabilise the anion inductively and by resonance	Most acidic as both NO_2 groups can stabilise the anion inductively and by resonance

1.7.2 Bases

A base is a substance that accepts a proton (Brønsted-Lowry). Basic compounds are good proton acceptors as the conjugate acids, formed on protonation, are relatively stable. Consequently, strong bases (B: or B$^-$) give conjugate acids (BH$^+$ or BH) with high pK_a values.

Equilibria and equilibrium
constants are discussed in
Section 4.9.1.1

In water:

basicity constant

$$\ddot{B} \quad + \quad H_2O \quad \overset{K_b}{\rightleftharpoons} \quad \overset{\oplus}{B}H \quad + \quad \overset{\ominus}{H}O$$

Base   Acid   Conjugate   Conjugate
                            acid   base

For the use of bases in elimination
reactions of halogenoalkanes, see
Section 5.3.2

The strength of bases is usually described by the K_a and pK_a values of the conjugate acid.

$$\overset{\oplus}{B}H \quad + \quad H_2O \quad \overset{K_a}{\rightleftharpoons} \quad \ddot{B} \quad + \quad H_3\overset{\oplus}{O}$$

For reactions of bases with
carbonyl compounds see Sections
8.4.3 and 9.11

$$K_a \approx \frac{[B][H_3\overset{\oplus}{O}]}{[\overset{\oplus}{B}H]}$$

As H_2O is in excess

- If B is a *strong* base then BH$^+$ will be relatively stable and not easily deprotonated. BH$^+$ will therefore have a *high* pK_a value.

Inductive effects are introduced
in Section 1.6.1

- If B is a *weak* base then BH$^+$ will be relatively unstable and easily deprotonated. BH$^+$ will therefore have a *low* pK_a value.

The cation can be stabilised by +I and +M groups, which can delocalise the positive charge. (The more 'spread out' the positive charge, the more stable it is.)

Mesomeric effects are introduced in Section 1.6.3

1.7.2.1 Inductive effects and aliphatic (or alkyl) amines

On protonation of amines (e.g. RNH_2), ammonium salts are formed.

Aliphatic amines have nitrogen bonded to one or more alkyl groups; aromatic amines have nitrogen bonded to one or more aryl groups

$$R-\overset{..}{N}H_2 \quad + \quad H^{\oplus} \quad \rightleftharpoons \quad R-\overset{\oplus}{N}H_3$$

primary amine   *ammonium ion*

The greater the +I effect of the R group, the greater the electron density at nitrogen and the more basic the amine. The greater the +I effect, the more stable the ammonium ion and the more basic the amine.

Primary (RNH_2), secondary (R_2NH) and tertiary (R_3N) amines are introduced in Section 2.1

no +I group   *three +I groups*

Triethylamine (Et_3N) is commonly used as a base in organic synthesis (Section 5.2.2)

pK_a 9.3	10.7	10.9	10.9

The pK_a values *should* increase steadily as more +I alkyl groups are introduced on nitrogen. However, the pK_a values are determined in *water*, and the more hydrogen atoms on the positively charged nitrogen, the greater the extent of hydrogen bonding between water and the cation. This solvation leads to the stabilisation of the cations containing N–H bonds.

Hydrogen bonds are introduced in Section 1.1

In organic solvents (which cannot solvate the cation) the order of pK_a's is expected to be as follows.

tertiary   *secondary*   *primary*
amine   *amine*   *amine*   *ammonia*

$$R_3\overset{..}{N} \quad > \quad R_2\overset{..}{N}H \quad > \quad R\overset{..}{N}H_2 \quad > \quad \overset{..}{N}H_3 \qquad (R = +I \text{ alkyl group})$$

most basic   *least basic*

The presence of −I and/or −M groups on nitrogen reduces the basicity and so, for example, primary amides ($RCONH_2$) are poor bases.

Secondary amides (RCONHR) and tertiary amides ($RCONR_2$) are also very weak bases because the nitrogen lone pairs are stabilised by resonance

Ethanamide

The C=O group stabilises the lone pair on nitrogen by resonance – this reduces the electron density on nitrogen

−M, −I

If ethanamide was protonated on nitrogen, the positive charge could not be stabilised by delocalisation. Protonation therefore occurs on oxygen as the charge can be delocalised on to the nitrogen atom.

Reactions of amides are discussed in Section 9.8

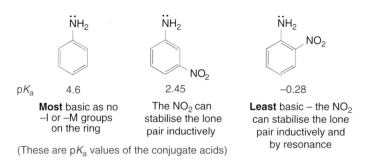

The conjugate acid has a low pK_a of −0.5

1.7.2.2　Mesomeric effects and aryl (or aromatic) amines

For the preparation and reactions of aniline (PhNH$_2$), see Section 7.8

The lone pair of electrons on the nitrogen atom of aminobenzene (or aniline, PhNH$_2$) can be stabilised by delocalisation of the electrons onto the 2-, 4- and 6-positions of the benzene ring. Aromatic amines are therefore less basic than aliphatic amines.

- If −M groups are introduced at the 2-, 4- and/or 6-positions (but not the 3- or 5-positions) the anion can be further stabilised by delocalisation, as the negative charge can be spread on to the −M group. This reduces the basicity of the amine.
- If −I groups are introduced on the benzene ring, the order of −I stabilisation is 2-position > 3-position > 4-position. This reduces the basicity of the amine.

pK_a　4.6	2.45	−0.28
Most basic as no −I or −M groups on the ring	The NO$_2$ can stabilise the lone pair inductively	**Least** basic – the NO$_2$ can stabilise the lone pair inductively and by resonance

(These are pK_a values of the conjugate acids)

For the Pauling electronegativity scale see Section 1.6.1

- If +M groups (e.g. OMe) are introduced at the 2-, 4- or 6-position of aminobenzene (PhNH$_2$), then the basicity is increased. This is because the +M group donates electron density to the carbon atom bearing the amine group. Note that the nitrogen atom, not the oxygen atom, is protonated – this is because nitrogen is less electronegative than oxygen and is a better electron donor.

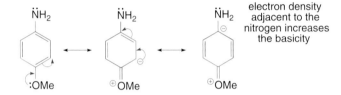

The OMe group is –I but +M	(These are pK_a values of the conjugate acids formed by protonation of the –NH_2 group)

The OMe group is called a methoxy group (see Section 2.4 for naming organic compounds)

pK_a	4.2	4.5	5.3
	Least basic as the OMe group cannot donate electron density to the carbon atom bearing the nitrogen	The OMe group can donate electron density to the nitrogen but it has a strong –I effect as it is in the 2-position	**Most** basic as the OMe group can donate electron density to the nitrogen and it has a weak –I effect (as well apart from the nitrogen)

Curly arrows can be used to show the delocalisation of electrons on to the carbon atom bearing the nitrogen.

Curly arrows are introduced in Section 4.1

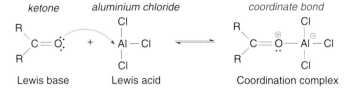

1.7.3 Lewis acids and bases

• A *Lewis acid* is any substance that accepts an electron pair in forming a coordinate bond (Section 1.1). Examples include H^+, BF_3, $AlCl_3$, $TiCl_4$, $ZnCl_2$ and $SnCl_4$. They have unfilled valence shells and so can accept electron pairs.

• A *Lewis base* is any substance that donates an electron pair in forming a coordinate bond. Examples include H_2O, ROH, RCHO, $R_2C=O$, R_3N and R_2S. They all have a lone pair(s) of electrons on the heteroatom (O, N or S).

A heteroatom is any atom that is not carbon or hydrogen

Reactions of ketones are discussed in Chapter 8

1.7.4 Basicity and hybridisation

The greater the 's' character of an orbital, the lower in energy the electrons and the more tightly the electrons are held to the nucleus. The electrons in an sp-orbital are therefore less available for protonation than those in an sp^2- or sp^3-orbital, and hence the compounds are less basic.

tertiary amine	*imine*	*nitrile*
most basic $R_3\ddot{N}$ >	$R_2C=\ddot{N}H$ >	$RC\equiv\ddot{N}$ **least** basic
alkyl anion	*alkenyl anion*	*alkynyl anion*
most basic $R_3C^{\ominus}$ >	$R_2C=\overset{\ominus}{C}H$ >	$RC\equiv\overset{\ominus}{C}$ **least** basic
sp^3	sp^2	sp
(25% s)	(33% s)	(50% s)

1.7.5 Acidity and aromaticity

Aromatic compounds are planar, conjugated systems which have $4n + 2$ electrons (Hückel's rule) (Section 7.1). If, on deprotonation, the anion is part of an aromatic π-system then the negative charge will be stabilised. Aromaticity will therefore *increase* the acidity of the compound.

Toluene is a common solvent. Oxidation of the CH_3 group is discussed in Section 7.6

fluorene
$pK_a = 22$

toluene
$pK_a = 40$

Resonance stabilisation of carbanions is introduced in Section 1.6.3

Each ring contributes 6π electrons

more stable

The anion is stabilised by resonance and it is aromatic (planar and 14π electrons)

less stable

Although the anion is stabilised by resonance it does not contribute to the aromaticity (this would give 8π electrons)

If a lone pair of electrons on a heteroatom is part of an aromatic π-system, then these electrons will not be available for protonation. Aromaticity will therefore *decrease* the basicity of the compound.

Reactions of aromatic heterocycles, including pyrrole and pyridine are discussed in Sections 7.10 and 7.11

Each double bond contributes 2π electrons

Pyrrole

The lone pair of electrons contributes to the 6π-electrons in the aromatic ring. Pyrrole is therefore not basic (pK_a −4)

Each double bond contributes 2π electrons

Pyridine

The lone pair of electrons does *not* contribute to the 6π-electrons in the aromatic ring. Pyridine is therefore basic (pK_a 5)

For a table of pK_a values see Appendix 3

1.7.6 Acid-base reactions

The pK_a values can be used to predict if an acid-base reaction can take place. An acid will donate a proton to the conjugate base of any acid with a higher pK_a value.

This means that the product acid and base will be more stable than the starting acid and base.

ethyne   *amide ion*   *ethynyl anion*   *ammonia*

$$HC\equiv C-H \ + \ {}^{\ominus}NH_2 \ \rightleftharpoons \ HC\equiv C^{\ominus} \ + \ NH_3$$

$pK_a\,25$

Ammonia has a higher pK_a value than ethyne $pK_a\,38$ and so the equilibrium lies to the right

Deprotonation of terminal alkynes is discussed in Section 6.3.2.5

propanone   *hydroxide ion*   *enolate ion*   *water*

$$CH_3COCH_3 \ + \ {}^{\ominus}OH \ \rightleftharpoons \ CH_3COCH_2^{\ominus} \ + \ H_2O$$

$pK_a\,20$

Water has a lower pK_a value than propanone $pK_a\,15.7$ and so the equilibrium lies to the left

For deprotonation of carbonyl compounds to form enolate ions, see Section 8.4.3

Worked Example

(a) Giving your reasons, rank the following carbanions **1–4** in order of increasing stability.

Hint: Determine whether the groups attached to the negatively charged carbons in **1–4** can stabilise the lone pair by I and/or M effects

1   2   3   4

(b) Identify, giving your reasons, the most acidic hydrogen atom(s) in compound **5**.

Hint: Consider a δ+ hydrogen atom bonded to an electronegative atom that, on deprotonation, gives the more stable conjugate base

5

6

Hint: Show all the lone pairs in **6** and consider their relative availability. Compare the stability of possible conjugate acids

(c) Identify, giving your reasons, the basic functional group in compound **6**.

Answer

(a)

3   2   4   1

increasing stability ⟶

Inductive and mesomeric effects (resonance) are discussed in Sections 1.6.1 and 1.6.3

For the *tert*-butyl anion **3**, because the three CH_3 are electron-donating groups (+I), this makes **3** less stable than the methyl anion **2**.

For the preparation and reactions of enolate ions, see Section 8.4.3

The benzyl anion **4** is more stable than the methyl anion **2** because it is stabilised by resonance – the negative charge is delocalised on to the 2, 4 and 6 positions of the ring.

Enolate ion **1** is the most stable because the anion is stabilised by resonance and one resonance form has the negative charge on oxygen – a negative charge on oxygen is more stable than a negative charge on carbon.

(b) Hydrogen atoms bonded to oxygen are more acidic than those bonded to carbon. As oxygen is more electronegative than carbon, the conjugate base is more stable. The carboxylic acid group is more acidic than the alcohol group in **5** because deprotonation of the carboxylic acid gives a conjugate base that is stabilised by resonance.

Formation of carboxylate ions is discussed in Section 1.7.1

(c) The tertiary amine is the most basic group in **6**. The lone pairs on the nitrogen atoms in the tertiary amide and aniline groups are both delocalised and less available for protonation (the oxygen atom of the tertiary amide is less basic than the tertiary amine because oxygen is more electronegative than nitrogen, hence the oxygen lone pairs are less available). On protonation of the tertiary amine, the conjugate acid is stabilised by three +I effects.

Functional groups are discussed in Section 2.1

Formation of ammonium ions is discussed in Section 1.7.2

Problems

Sections 1.6.1 and 1.6.3

1. Using the I and M notations, identify the electronic effects of the following substituents.

(a) −Me (b) −Cl   (c) −NH$_2$ (d) −OH
(e) −Br (f) −CO$_2$Me (g) −NO$_2$ (h) −CN

2. (a) Use curly arrows to show how cations **A**, **B** and **C** (shown below) are stabilised by resonance, and draw the alternative resonance structure(s).   Section 1.6.3

      A   **B**   **C**

   (b) Would you expect **A**, **B** or **C** to be the more stable? Briefly explain your reasoning.

3. Provide explanations for the following statements.   Sections 1.6.3 and 1.7.1

   (a) The carbocation CH$_3$OCH$_2^+$ is more stable than CH$_3$CH$_2^+$.
   (b) 4-Nitrophenol is a much stronger acid than phenol (C$_6$H$_5$OH).
   (c) The pK_a of CH$_3$COCH$_3$ is much lower than that of CH$_3$CH$_3$.
   (d) The C−C single bond in CH$_3$CN is longer than that in CH$_2$=CH−CN.
   (e) The cation CH$_2$=CH−CH$_2^+$ is resonance stabilised whereas the cation CH$_2$=CH−NMe$_3^+$ is not.

4. Why is cyclopentadiene (pK_a 15.5) a stronger acid than cycloheptatriene (p$K_a \sim 36$)?   Section 1.7.5

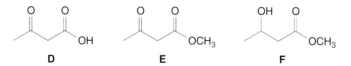

      cyclopentadiene   cycloheptatriene

5. Which hydrogen atom would you expect to be the most acidic in each of the following compounds?   Section 1.7.1

   (a) 4-Methylphenol (or *p*-cresol, 4-HOC$_6$H$_4$CH$_3$)
   (b) 4-Hydroxybenzoic acid (4-HOC$_6$H$_4$CO$_2$H)
   (c) H$_2$C=CHCH$_2$CH$_2$C≡CH
   (d) HOCH$_2$CH$_2$CH$_2$C≡CH

6. Arrange the following sets of compounds in order of decreasing basicity. Briefly explain your reasoning.   Section 1.7.2

   (a) 1-Aminopropane, ethanamide (CH$_3$CONH$_2$), guanidine [HN=C(NH$_2$)$_2$], aniline (C$_6$H$_5$NH$_2$).
   (b) Aniline (C$_6$H$_5$NH$_2$), 4-nitroaniline, 4-methoxyaniline, 4-methylaniline.

7. For each of the following compounds **D**−**F**, identify the most acidic hydrogen atom(s). Briefly explain your reasoning.   Section 1.7.1

       D   **E**   **F**

Section 1.7.2

8. For each of the following compounds **G–I**, identify the most basic group. Briefly explain your reasoning.

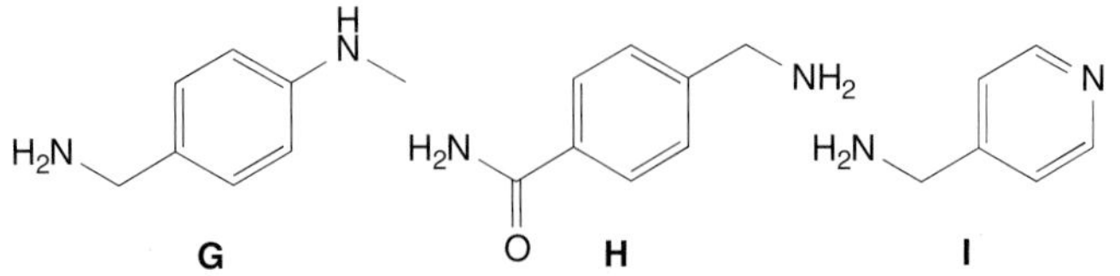

G   O   H   I

Section 1.7.6

9. Given the approximate pK_a values shown below, for the following acid-base reactions (a)–(e), determine whether the position of the equilibrium lies over to the reactant side or the product side.

Acid	pK_a value
PhOH	9.9
H_2O	15.7
CH_3COCH_3	20
H_2	35
NH_3	38
$H_2C=CH_2$	44

(a) $NaH + PhOH \rightleftharpoons PhO^{\ominus} Na^{\oplus} + H_2$

(b) $CH_3COCH_3 + NaOH \rightleftharpoons CH_3COCH_2^{\ominus}Na^{\oplus} + H_2O$

(c) $H_2C=CH_2 + NaNH_2 \rightleftharpoons H_2C=CH^{\ominus} Na^{\oplus} + NH_3$

(d) $CH_3COCH_2^{\ominus} Na^{\oplus} + PhOH \rightleftharpoons CH_3COCH_3 + PhO^{\ominus}Na^{\oplus}$

(e) $H_2C=CH^{\ominus} Na^{\oplus} + H_2O \rightleftharpoons H_2C=CH_2 + NaOH$

2

Functional groups, nomenclature and drawing organic compounds

Key point. Organic compounds are classified by *functional groups*, which determine their chemistry. The names of organic compounds are derived from the functional group (or groups) and the main carbon chain. From the name, the structure of organic compounds can be drawn using *full structural formulae*, *condensed structural formulae* or *skeletal* structures.

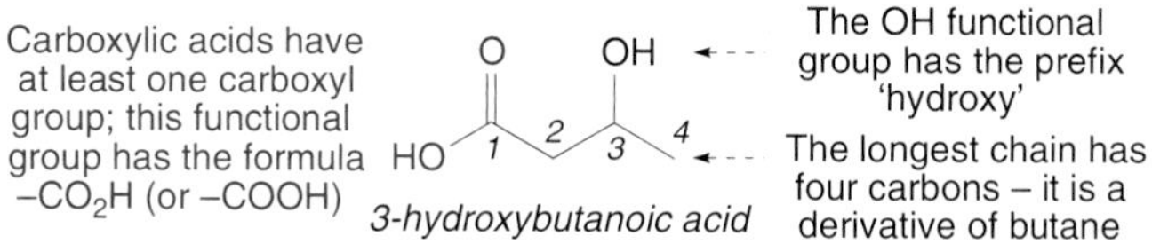

Carboxylic acids have at least one carboxyl group; this functional group has the formula $-CO_2H$ (or $-COOH$)

The OH functional group has the prefix 'hydroxy'

The longest chain has four carbons – it is a derivative of butane

3-hydroxybutanoic acid

2.1 Functional groups

A functional group is made up of an atom or atoms with characteristic chemical properties. The chemistry of organic compounds is determined by the functional groups that are present.

Hydrocarbons (only hydrogen and carbon are present)

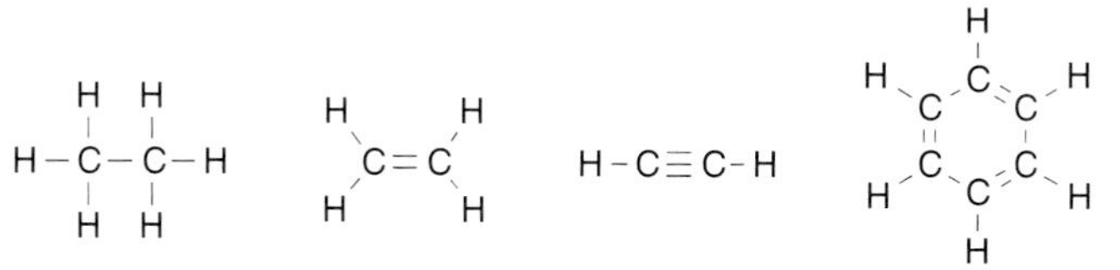

ethane
(an alkane)
single CC bond

ethene
(an alkene)
double CC bond

ethyne
(an alkyne)
triple CC bond

benzene
(an arene)
single/double
CC bonds

The common name for ethyne is acetylene

The lengths of all six CC bonds in benzene are between those for a C–C and a C=C bond (Section 7.1)

Keynotes in Organic Chemistry, Second Edition. Andrew F. Parsons.
© 2014 John Wiley & Sons, Ltd. Published 2014 by John Wiley & Sons, Ltd.

Alkanes are saturated as they contain the maximum number of hydrogen atoms per carbon (C_nH_{2n+2}). Alkenes, alkynes and arenes are all unsaturated.

Carbon bonded to an electronegative atom(s)

- single bond (R is the carbon framework, typically an alkyl group; see Section 2.2)

The structure of the nitro group in a nitro compound:

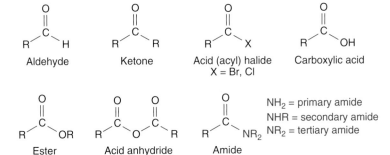

Halogenoalkanes (X = F, Cl, Br, I)	
RCH_2X	primary (1°)
R_2CHX	secondary (2°)
R_3CX	tertiary (3°)

Alcohols	
RCH_2OH	primary alcohol
R_2CHOH	secondary alcohol
R_3COH	tertiary alcohol

$R-O-R$ Ether $R-NO_2$ Nitro compound

$R-SH$ Thiol $R-S-R$ Sulfide (thioether)

Amines	
$R-NH_2$	primary amine
$R-NHR$	secondary amine
$R-NR_2$	tertiary amine
$R-\overset{\oplus}{N}R_3$	quaternary ammonium ion

- double bond to oxygen (these are called carbonyl compounds)

Cyclic esters are called lactones. Cyclic amides are called lactams.

Aldehyde   Ketone   Acid (acyl) halide X = Br, Cl   Carboxylic acid

If the two R groups in an acid anhydride are the same, it is called a symmetrical anhydride

Ester   Acid anhydride   Amide

NH_2 = primary amide
NHR = secondary amide
NR_2 = tertiary amide

- triple bond to nitrogen

$$R-C\equiv N$$
Nitrile

2.2 Alkyl and aryl groups

When a hydrogen atom is removed from an alkane this gives an alkyl group. The symbol R is used to represent a general alkyl group (i.e. a methyl, ethyl, propyl, etc. group).

The IUPAC name for isopropyl is 1-methylethyl

The IUPAC name for *tert*-butyl is 1,1-dimethylethyl

name (symbol)	structure	name (symbol)	structure
methyl (Me)	$-CH_3$	isopropyl (^iPr)	$-CH(CH_3)_2$
ethyl (Et)	$-CH_2CH_3$	isobutyl (^iBu)	$-CH_2CH(CH_3)_2$
propyl (Pr)	$-CH_2CH_2CH_3$	sec-butyl (^sBu)	$-CH(CH_3)CH_2CH_3$
butyl (Bu)	$-CH_2CH_2CH_2CH_3$	tert-butyl (^tBu)	$-C(CH_3)_3$

When a hydrogen atom is removed from a benzene ring this gives a phenyl group (Ph). Related groups include the benzyl group (PhCH$_2$).

Do not confuse Ph with pH (Section 1.7) or with phenol, C_6H_5OH

phenyl (C_6H_5), Ph

aryl, Ar
X = various
functional group(s)

benzyl (PhCH$_2$), Bn

vinyl

allyl

the wavy line indicates where the group is joined to the rest of the structure

It is useful to draw a benzene ring as alternating C=C and C–C bonds as this helps to keep track of electron movement in reaction mechanisms (Section 7.2)

2.3 Alkyl substitution

A primary (or 1°) carbon is bonded to one other carbon

A secondary (or 2°) carbon is bonded to two other carbons

A tertiary (or 3°) carbon is bonded to three other carbons

A quaternary (or 4°) carbon is bonded to four other carbons

2.4 Naming carbon chains

The IUPAC name of an organic compound is composed of three parts.

Functional groups are introduced in Section 2.1

Prefix — Parent — Suffix

What substituents (e.g. *minor* functional groups) are on the main chain and where are they?

What is the length of the main carbon chain?

What is the *major* functional group?

There are *four* key steps in naming organic compounds.

1. Find the longest carbon chain and name this as an alkane. This is the parent name.

No. of Carbons	Alkane Name	No. of Carbons	Alkane Name
1	methane	6	hexane
2	ethane	7	heptane
3	propane	8	octane
4	butane	9	nonane
5	pentane	10	decane

2. Identify the major functional group. Replace -ane (in the alkane) with a suffix.

> **Functional group priorities**
>
> carboxylic acid (RCO_2H) > ester (RCO_2R) > acid (acyl) chloride ($RCOCl$) > amide ($RCONH_2$) > nitrile (RCN) > aldehyde ($RCHO$) > ketone ($RCOR$) > alcohol (ROH) > amine (RNH_2) > alkene ($RCH=CHR$) > alkane (RH) > ether (ROR) > halogenoalkane (RX)

major functional group	suffix	major functional group	suffix
alkene	−ene	aldehyde	−al
alkyne	−yne	ester	−oate
alcohol	−ol	ketone	−one
amine	−amine	carboxylic acid	−oic acid
nitrile	−nitrile	acid (acyl) chloride	−oyl chloride

A branch point is where a carbon atom forms bonds to three or four carbon atoms

3. Number the atoms in the main chain. Begin at the end nearer the major functional group and give this the lowest number. For alkanes, begin at the end nearer the first branch point.
4. Identify the substituents (e.g. minor functional groups) on the main chain and their number. Two substituents on the same carbon are given the same number. The substituent name and position is the prefix. The names of two or more different substituents should be included in alphabetical order in the prefix (e.g. hydroxy before methyl).

minor functional group	prefix	minor functional group	prefix
chloride	chloro−	aldehyde	formyl−
bromide	bromo−	ketone	oxo−
iodide	iodo−	alkane	alkyl−
alcohol	hydroxy−	alkene	alkenyl−
ether	alkoxy−	nitro	nitro−
amine	amino−	nitrile	cyano−

Di- or tri- is used in the prefix or suffix to indicate the presence of two or three of the minor or major functional groups (or substituents), respectively.

Examples

For alcohols, the position of the OH group is sometimes shown at the front of the name of the parent alkane, e.g. 2-propanol

For ketones, the position of the C=O bond is sometimes shown at the front of the name of the parent alkane, e.g. 2-pentanone

3-methylpentanoic acid

1,2-dichloropropane

1-aminopropan-2-ol

4-hydroxy-4-methylpentan-2-one

2.4.1 Special cases

2.4.1.1 Alkenes and alkynes

The position of the double or triple bond is indicated by the number of the lowest carbon atom in the alkene or alkyne.

4-methylpent-2-ene

2-methylbuta-1,3-diene
(an example of a conjugated diene: two C=C bonds separated by one C–C bond)

major functional group - - - - → HO

2-methyl-3-butyn-2-ol

For alkenes, the position of the C=C bond is sometimes shown at the front of the name, e.g. 2-pentene

2.4.1.2 Aromatics

Monosubstituted benzene derivatives are usually named after benzene (C_6H_6), although some non-systematic or common names (in brackets) are still used.

Reactions of benzene and substituted benzenes are discussed in Chapter 7

X	Name	X	Name
H	benzene	CH_3	methylbenzene (toluene)
Br	bromobenzene	$CH=CH_2$	ethenylbenzene (styrene)
Cl	chlorobenzene	OH	hydroxybenzene (phenol)
NO_2	nitrobenzene	NH_2	aminobenzene (aniline)
		CN	cyanobenzene (benzonitrile)

The word benzene comes first when functional groups of higher priority (than benzene) are on the ring

X	Name
CHO	benzenecarboxaldehyde (benzaldehyde)
CO_2H	benzenecarboxylic acid (benzoic acid)

Disubstituted derivatives are sometimes named using the prefixes *ortho-* (or positions 2- and 6-), *meta-* (or positions 3- and 5-) and *para-* (or position 4-).

For trisubstituted derivatives, the lowest possible numbers are used and the prefixes are arranged alphabetically.

ortho(o) 6 — 2 *ortho(o)*
meta(m) 5 — 3 *meta(m)*
4 *para(p)*
1 *ipso*

p–bromophenol

3-chloro-4-hydroxy-benzoic acid

2,4-dinitrotoluene

Aromatic compounds that contain at least one heteroatom (e.g. O, N or S) as part of the ring are called aromatic heterocyclic compounds (heterocycles).

Reactions of aromatic heterocycles are discussed in Sections 7.10 and 7.11

pyridine
(a base, commonly
used in synthesis)

pyrrole   furan   2,4-dimethylfuran

2.4.2.3 Esters

These are named in two parts. The first part represents the R^1 group attached to oxygen. The second represents the R^2CO_2 portion which is named as an alkanoate (i.e. the suffix is –oate). A space separates the two parts of the name.

The preparation of esters is discussed in Section 9.4.2 and their reactions in Section 9.7

methyl propanoate

ethyl benzoate

2.4.2.4 Amides

The R^1 group is the prefix, and $N–$ is written before this to show the group is attached to nitrogen.

Reactions of amides are described in Section 9.8

N-methylpropanamide
(a secondary amide)

Primary, secondary and tertiary amides are introduced in Section 2.1

2.4.2.5 Cyclic non-aromatic compounds

Alicyclic compounds are cyclic compounds that are not aromatic. (In contrast, *aliphatic* compounds are non-cyclic compounds that have an open chain of atoms.)

Cyclic compounds that have a ring of carbon atoms, which are not aromatic, are named using the prefix cyclo. For example, cyclobutane is a cyclic alkane with four carbon atoms. The atoms in the ring are numbered so that the smallest numbers indicate the position of substituents.

Conformations of cycloalkanes (cyclic alkanes) are discussed in Sections 3.2.3 and 3.2.4

cyclohexanol   3-bromocyclopent-1-ene   2-methylcyclopentanone

Cyclic compounds containing at least one heteroatom (e.g. O, N or S) are examples of heterocyclic compounds (heterocycles).

For epoxide formation and hydrolysis see Section 6.2.2.6

ethylene oxide
(cyclic ether called an epoxide or oxirane)

tetrahydrofuran
(a cyclic ether; a common solvent)

β-propiolactam
(a cyclic amide or lactam)

the carbon adjacent to a C=O bond is called the α-carbon

α-Substitution reactions of carbonyl compounds are discussed in Sections 8.4 and 9.10

2.5 Drawing organic structures

- In *full structural formulae*, every carbon atom and every C–H bond are shown.
- In *condensed* structures, the C–H bonds, and often the C–C bonds, are omitted.
- In *skeletal* structures the carbon and hydrogen atoms are not shown and the bonds to hydrogen are usually also not shown (although hydrogen atoms within functional groups, e.g. alcohols, amines, aldehydes and carboxylic acids, are shown). All other atoms are written. These structures are the most useful (and recommended for use by the reader) because they are uncluttered and quickly drawn, whilst showing all of the important parts of the molecule. Skeletal structures are usually drawn to indicate the approximate shape of the molecule, which is determined by the hybridisation of the atoms (Section 1.5).

Carbon chains with C–C or C=C bonds have an approximate zigzag shape (Section 3.2.2)

methyl propanoate

$$H-C-C-C-O-C-H \equiv CH_3CH_2CO_2CH_3 \equiv$$

full structural formula   condensed   skeletal

(E)-but-2-en-1-ol

$$H-C-C=C-C-OH \equiv CH_3CH=CHCH_2OH \equiv$$

full structural formula   condensed   skeletal

Alkenes like but-2-en-1-ol can exist as two configurational isomers; these isomers are named *cis-* or *trans-*, or *Z-* or *E-* (Section 3.3.1)

Benzene can be written with a circle within the ring to show the delocalisation of electrons (Section 7.1). However, this does not show the 6 π-electrons, which makes drawing reaction mechanisms impossible. A single resonance form, showing the three C=C bonds, is therefore most often used.

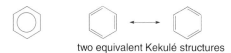

two equivalent Kekulé structures

Worked example

The following questions relate to the synthesis of aspirin, which is shown below.

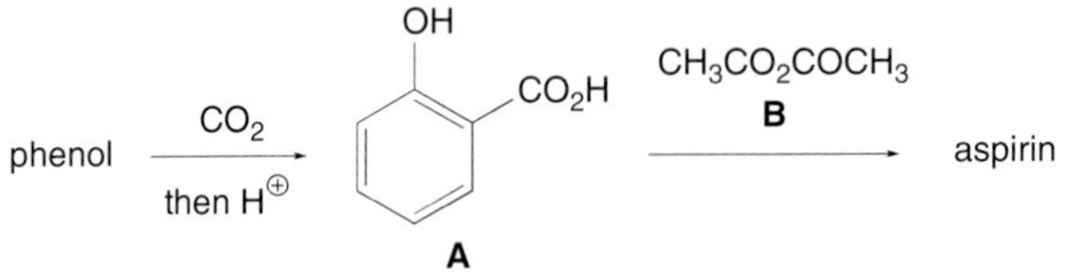

(a) Draw the skeletal structure of phenol.

(b) Give the IUPAC name for compound **A**.

(c) Draw the skeletal structure for compound **B**.

(d) Given that the IUPAC name of aspirin is 2-ethanoyloxybenzoic acid, draw the skeletal structure of aspirin.

(e) Draw the skeletal structure of *p*-ethanoyloxybenzoic acid.

Hint: Compound **A** is a derivative of benzoic acid

Hint: Compound **B** is an acid anhydride

Answer

(a)

Phenol is introduced in Section 2.4

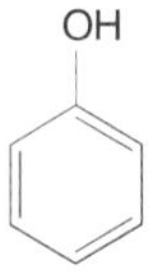

(b) 2-Hydroxybenzoic acid (the carboxylic acid group has a higher priority than an OH group)

Naming substituted benzenes introduced in Section 2.4

Acid anhydrides are introduced in Section 2.1 and skeletal structures in Section 2.5

(c)

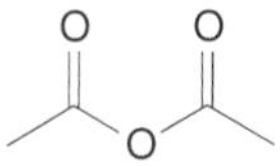

(d)

Naming substituted benzenes is introduced in Section 2.4 and skeletal structures in Section 2.5

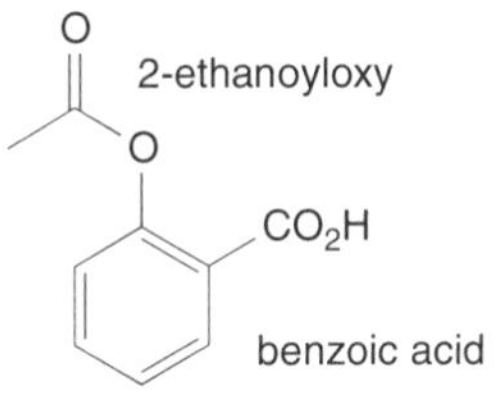

(e)

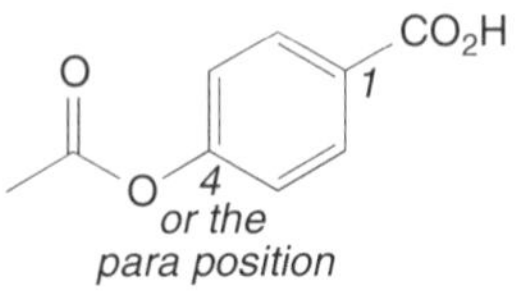

Problems

1. Draw skeletal structures for each of the following compounds.   Sections 2.4 and 2.5

   (a) 1-Bromo-4-chloro-2-nitrobenzene
   (b) Methyl 3-bromobutanoate
   (c) *N*-Methylphenylethanamide
   (d) 2-(3-Oxobutyl)cyclohexanone
   (e) Hex-4-en-2-one
   (f) 2-Buten-1-ol
   (g) 6-Chloro-2,3-dimethylhex-2-ene
   (h) 1,2,3-Trimethoxypropane
   (i) 2,3-Dihydroxybutanedioic acid (tartaric acid)
   (j) 5-Methylhex-4-enal

2. Name the following compounds.   Sections 2.4 and 2.5

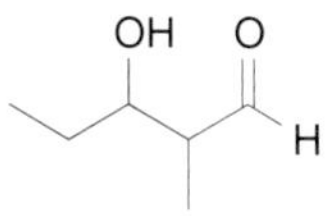

    (a)    (b) 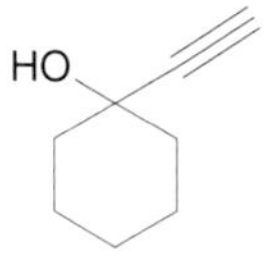   (c)

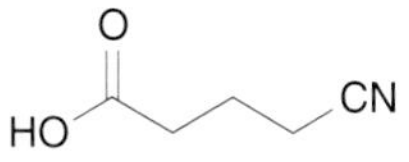

    (d) 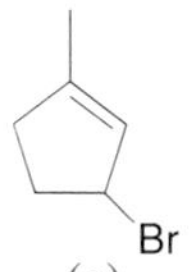   (e) 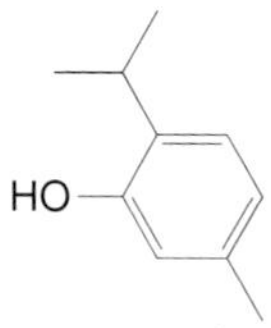   (f)

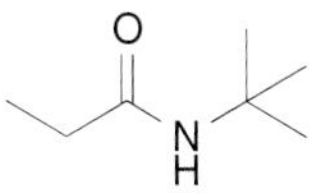

    (g) 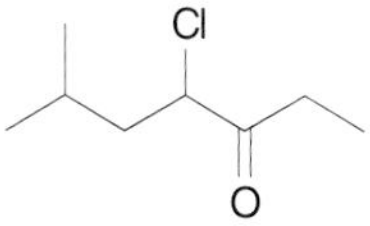   (h)

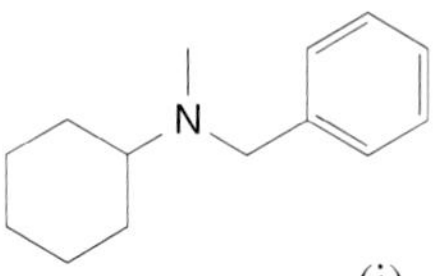

    (i) 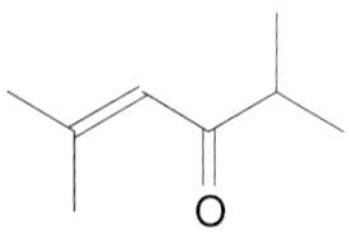   (j)

3. The following questions are based on the synthesis of the medicine ibuprofen, which is shown below.

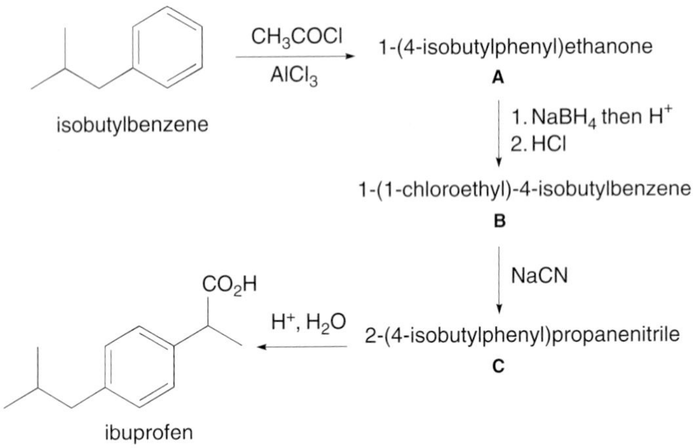

Section 2.2
Sections 2.4 and 2.5

(a) Label each carbon atom in the isobutyl group as either a primary, secondary, tertiary or quaternary carbon atom.
(b) Give the IUPAC name of CH_3COCl.
(c) Draw skeletal structures for compounds **A**, **B** and **C**.
(d) Give the IUPAC name of ibuprofen.

4. Chlorphenamine **G** is an antihistamine used in the treatment of allergic conditions. It can be prepared using the 3-step approach shown below.

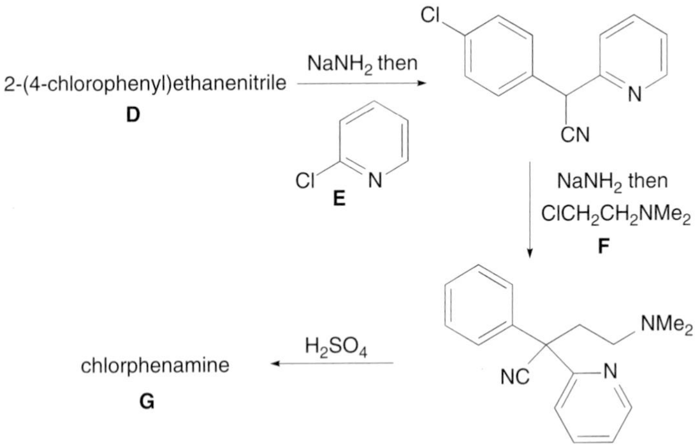

Section 2.5
Section 2.4
Sections 2.5 and 2.1

(a) Draw a skeletal structure for compound **D**.
(b) Give the IUPAC name of **E**.
(c) Draw a skeletal structure for compound **F**. Is **F** a primary, secondary or tertiary amine?

Section 2.4

(d) Given that the IUPAC name of chlorphenamine (**G**) is *N,N*-dimethyl-3-phenyl-3-(pyridin-2-yl)propan-1-amine), draw a skeletal structure of **G**.

3

Stereochemistry

Key point. The spatial arrangement of atoms determines the *stereochemistry*, and hence the shape, of organic molecules. When different shapes of the same molecule are interconvertable on rotating a bond, they are called *conformational isomers*. In contrast, *configurational isomers* cannot be interconverted without breaking a bond and examples include alkenes and *isomers with chiral centres*, which rotate plane-polarised light.

Newman projection of the staggered conformation of ethane (CH_3–CH_3)

The *E*–isomer of but-2-enoic acid

The *R*–isomer of 2-hydroxypropanoic acid (D–(–)-lactic acid)

3.1 Isomerism

Isomers are compounds that have the same numbers and kinds of atoms, but they differ in the way that the atoms are arranged. The presence of different isomers is called **isomerism**.

Isomers may have different carbon skeletons, different functional groups, or the same functional groups at different positions

- *Structural (constitutional) isomers* are compounds which have the same molecular formula, but have the atoms joined together in a different way. They have different physical and chemical properties.

Keynotes in Organic Chemistry, Second Edition. Andrew F. Parsons.
© 2014 John Wiley & Sons, Ltd. Published 2014 by John Wiley & Sons, Ltd.

Examples: chain isomers, position isomers and functional group isomers

Naming organic compounds is introduced in Section 2.4

$$
\begin{array}{c}
\quad CH_3 \\
\quad | \\
H_3C-C-CH_3 \\
\quad | \\
\quad H
\end{array}
\quad \text{and} \quad H_3C-CH_2-CH_2-CH_3
$$

2-methylpropane
(branched chain)

butane
(straight chain)

Both C_4H_{10} but different carbon skeleton: chain isomers

Functional groups are discussed in Section 2.1

Position isomers are often called regioisomers (Section 4.8)

$$
\begin{array}{c}
\quad OH \\
\quad | \\
H_3C-C-CH_3 \\
\quad | \\
\quad H
\end{array}
\quad \text{and} \quad H_3C-CH_2-CH_2-OH
$$

propan-2-ol
(secondary alcohol)

propan-1-ol
(primary alcohol)

Both C_3H_8O but different position of the functional group: position isomers

Diethyl ether is a common solvent in organic synthesis (Section 8.3.4.2)

$$
\begin{array}{c}
\quad OH \\
\quad | \\
H_3C-C-CH_3 \\
\quad | \\
\quad CH_3
\end{array}
\quad \text{and} \quad H_3C-CH_2-O-CH_2-CH_3
$$

2-methylpropan-2-ol
(tertiary alcohol)

diethyl ether
(ether)

Both $C_4H_{10}O$ but different functional groups: functional group isomers

Conformational isomers are sometimes called rotamers

- *Conformational isomers* (*conformers*) are different shapes of the same molecule resulting from rotation around a single bond, such as C–C. They are not different compounds (i.e. they have the same physical and chemical properties) and are readily interconvertible (Section 3.2).
- *Configurational isomers* have the same molecular formula and, although the atoms are joined together in the same way, they are arranged differently in space (with respect to each other). They are not readily interconvertible (Section 3.3).

3.2 Conformational isomers

The different arrangements of atoms caused by rotation about a single bond are called conformations. A conformational isomer, or *conformer*, is a compound with a particular conformation. Conformational isomers can be represented by *Sawhorse projections* or *Newman projections*.

3.2.1 Conformations of ethane (CH_3CH_3)

Rotation about the C–C bond produces two distinctive conformations.

- *Eclipsed conformation*–C–H bonds on each carbon atom are as close as possible.

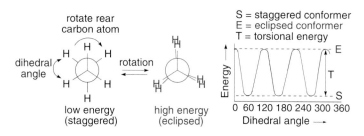

Staggered conformation – C–H bonds on each carbon atom are as far apart as possible.

The staggered conformation is more stable as the C–H bonds are further apart (the pairs of electrons in the bonds repel on another). The energy difference between them ($\sim$12 kJ mol^{-1}) is known as the *torsional strain*. This energy difference (or *energy barrier to rotation*) is relatively small so there is free rotation about the C–C bond at room temperature.

Torsional strain is often called torsional energy

The angle between the C–H bonds on the front and back carbon is known as the *dihedral* (or *torsional*) *angle*.

3.2.2 Conformations of butane ($CH_3CH_2CH_2CH_3$)

In butane, the staggered conformations do not all have the same energy.

- The *anti-periplanar conformation* is the most stable as the two methyl groups are as far apart as possible (180° separation). Most butane molecules will adopt this shape for most of the time.

For E2 elimination reactions, halogenoalkanes need to adopt an anti-periplanar conformation (Section 5.3.2.1)

- The *synclinal* (*gauche*) *conformation* is higher in energy (by $4\,kJ\,mol^{-1}$) as the two methyl groups are near one another (60° separation) resulting in *steric strain* (*steric hindrance*). Steric strain is the repulsive interaction between two groups which are closer to one another than their atomic radii allow.
- The *anticlinal conformation* is higher in energy (by $16\,kJ\,mol^{-1}$) as a hydrogen atom and methyl group are forced extremely close to one another (0° separation).
- The *syn-periplanar conformation* is the least stable (it is higher in energy by $19\,kJ\,mol^{-1}$) as the two methyl groups are forced extremely close to one another (0° separation).

For Ei eliminations, the starting materials adopt a syn-periplanar conformation (Section 6.2.1)

For a discussion of the importance of steric effects in reactions, see Section 4.4

The most stable conformation for a straight-chain alkane is a zigzag shape as the alkyl groups are as far apart as possible. (This is why we draw alkyl chains with a zigzag structure.)

For naming carbon chains, see Section 2.4

zigzag shape of butane   zigzag shape of hexane

3.2.3 Conformations of cycloalkanes

The shape of cycloalkanes is determined by torsional strain, steric strain and angle strain.

Hybridisation is discussed in Section 1.5

- *Angle strain.* For an sp^3 hybridised carbon atom the ideal bond angle is 109.5° (tetrahedral). Angle strain is the extra energy that a compound has because of non-ideal bond angles (i.e. angles above or below 109.5°).

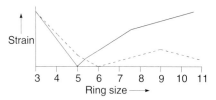

— = **Angle strain**

A 3-ring has the highest and a planar 5-ring has the lowest angle strain. After 5, the angle strain increases as the ring gets larger

- - - - = **Total strain**
(angle + steric + torsional strain)
A 3-ring has the highest total strain which reaches a minimum for a 6-ring. The strain increases from 6 to 9 and then decreases

Cyclopentane (5-membered) and cyclohexane (6-membered) rings are therefore the most stable and, consequently, the most easily formed.

Cycloalkanes can adopt different conformations (or shapes): cyclopropane is flat, cyclobutane forms a butterfly shape while cyclopentane forms an envelope shape.

For naming cycloalkanes, see Section 2.4

Cyclopropanes can be formed by addition of carbenes to alkenes (Section 6.2.2.9)

cyclopropane
(internal C–C
angle = 60°)

cyclobutane (butterfly)
(internal C–C angle = 88°)

cyclopentane (envelope)
(internal C–C angle = 105°)

Steroids contain a cyclopentane ring linked to three cyclohexane rings (Section 11.2.2)

Cyclopropane has to be planar and therefore has very strained bond angles of 60° and a great deal of torsional energy. Cyclobutane and cyclopentane can adopt non-planar (puckered) shapes that decrease the torsional strain by staggering the C–H bonds. However, this is at the expense of angle strain and the butterfly and envelope shapes represent the best compromise between the two opposing effects.

3.2.4 Cyclohexane

Cyclohexane adopts the chair or boat conformations, which are both free of angle strain. However, the boat conformation is less stable because of steric strain between the C-1 and C-4 (or flagstaff) hydrogens. The two chair forms can interconvert via the boat form in a process known as *ring-flipping*.

Chair
(strain-free)

steric strain
Boat

Chair
(strain-free)

Cyclohexanes can be formed by
hydrogenation of benzene (Section
7.7)

Newman projections

Chair   Boat   Chair

The chair conformation has six *axial* and six *equatorial* hydrogens. On ring-flipping, the axial hydrogens become equatorial and the equatorial hydrogens become axial.

For elimination reactions of
halocyclohexanes, requiring axial
C–H and C–X bonds, see
Section 5.3.2.1

H_a = axial hydrogen (point up and down)

H_e = equatorial hydrogen (point sideways)

If a substituent (X) is present, then this prefers to sit in an *equatorial position*. The equatorial conformer is lower in energy because *steric strain* (or *1,3-diaxial interactions*) raises the energy of the axial conformer. As the size of the X group increases, so does the proportion of the equatorial conformer at equilibrium.

X	% eq. : % ax.	
Me	95	5
^{t}Bu	>99	<1

axial
conformer

equatorial
conformer

1,3-diaxial
interactions
(unfavourable)

Steroids contain three cyclohexane
rings linked to a cyclopentane ring
(Section 11.2.2)

For disubstituted cyclohexanes, both groups should sit in an equatorial position. When this is not possible, then the largest group (e.g. a ^{t}Bu rather than a Me group) will sit in the equatorial position.

diaxial
conformer
(higher energy)

diequatorial
conformer
(lower energy)

The bulkier ^{t}Bu
group prefers the
equatorial position

When the two substituents on the ring are *both* pointing up (or *both* pointing down), then these cyclic compounds are designated as *cis-* stereoisomers. When one substituent is pointing up and the other down, then these cyclic compounds are designated as *trans-* stereoisomers (Section 3.3).

Naming cycloalkanes is discussed
in Section 2.4

cis-1-bromo-2-methylcyclohexane   *trans*-1-bromo-2-methylcyclohexane

3.3 Configurational isomers

The spatial arrangement of atoms or groups in molecules is known as *configuration*. Compounds with the same molecular formula and bonds, which cannot be interconverted without breaking a bond, have different configurations (and are known as configurational isomers).

3.3.1 Alkenes

Alkenes with two different substituents (A, B, D or E) at each end of the double bond can exist as two configurational isomers because there is no rotation around the C=C bond.

An alkene contains a C=C bond (Section 2.1)

can have configurational isomers

3.3.1.1 Cis- and trans- *isomerism*

For disubstituted alkenes (with two substituents on the double bond), alkenes can be named using the *cis-trans* nomenclature:

Cis- and trans- are also used to indicate the position of groups on rings that are joined together (Section 11.2.2)

- *cis*-isomers have the substituents on the *same* side
- *trans*-isomers have the substituents on the *opposite* side

Naming alkenes is discussed in Section 2.4

    Cis- and *trans*-isomers are also called geometric isomers because they have different shapes.

    There are no rigid rules for deciding whether a variety of substituted double bonds are *cis-* or *trans-* and so the more systematic *E/Z* nomenclature is generally preferred (Section 3.3.1.2).

3.3.1.2 E and Z nomenclature

For di-, tri- and tetra-substituted alkenes (with two, three and four substituents, respectively, on the double bond), alkenes can be named using the *E/Z* nomenclature. The groups on the double bond are assigned priorities based on a series of sequence rules.

- *E*-alkenes have the groups of highest priority on the *opposite* sides
- *Z*-alkenes have the groups of highest priority on the *same* sides

E isomers correspond to trans- isomers and Z isomers correspond to cis- isomers

Sequence rules

1. Rank the atoms directly attached to the double bond carbons in order of decreasing atomic number. The highest atomic number is ranked first (e.g. Br has a higher priority than Cl).

$$Atom = Br > Cl > O > N > C > H$$

$$Atomic\ number = 35 > 17 > 8 > 7 > 6 > 1$$

2. If the atoms directly linked to the double bond are the same then the second, third, fourth, etc. atoms (away from the double bond) are ranked until a difference is found.

Naming substituents is discussed in Section 2.4

$$\zeta-CH_2-CH_3 \quad > \quad \zeta-CH_3 \qquad \zeta-O-CH_3 \quad > \quad \zeta-OH$$

$$\quad\ \ ethyl \qquad\qquad\quad methyl \qquad\qquad methoxy \qquad\quad hydroxy$$

3. Multiple (double or triple) bonds are assumed to have the same number of single-bonded atoms.

Functional groups are introduced in Section 2.1

Assume C is bonded to 2 oxygens   Assume C is bonded to 3 nitrogens
Assume O is bonded to 2 carbons   Assume N is bonded to 3 carbons

Examples (numbers on each carbon are given in italic to indicate priorities)

(Z)-2-chloro-3-methylpent-2-ene

(E)-2-bromo-3-hydroxymethyl-pent-2-enenitrile

(E)-3-methyl-4-phenylpent-3-en-2-one

(Z)-3-hydroxymethyl-4-oxo-2-phenylbut-2-enoic acid

E or Z is shown, in brackets, at the front of the name

3.3.2   Isomers with chiral centres

- *Isomers with chiral centres (optical isomers)* are configurational isomers with the same chemical and physical properties, which are able to rotate plane-polarised light clockwise or anticlockwise.
- Asymmetrical molecules, which are non-identical with their mirror images, are known as *chiral* molecules and these can rotate plane-polarised light. If a molecule has a plane of symmetry it cannot be chiral and it is known as an *achiral* molecule.

- Molecules with a (asymmetric) tetrahedral carbon atom bearing four different groups are chiral. The 3-dimensional structure of the molecule, showing the position of the groups attached to the tetrahedral carbon atom, can be represented using solid/dashed wedges in *hashed-wedged line notation* (*flying wedge formulae*).

wedged line, the bond points toward you

hashed line, the bond points away from you

single line, shows the (two) bonds in the plane of the paper

3.3.2.1   Enantiomers

The tetrahedral asymmetric carbon atom is known as a *chiral* or *stereogenic centre*. These molecules, which are non-identical (not superimposable) with their mirror images, are called *enantiomers*.

Enantiomers of natural amino acids are discussed in Section 11.3

When $R^1 \neq R^2 \neq R^3 \neq R^4$, then these two mirror images are not superimposable–they are **enantiomers**

Enantiomers rotate plane-polarised light in opposite directions. The (+)-*enantiomer* (or dextrorotatory enantiomer) rotates the light to the right while the (−)-*enantiomer* (or laevorotatory enantiomer) rotates the light to the left. The amount of rotation is called the *specific rotation*, [α], and this is measured using a polarimeter. (By convention, the units of [α], 10^{-1} degree $cm^2 g^{-1}$, are often not quoted.)

$$[\alpha]_D^T = \frac{\text{observed rotation (degrees)} \times 100}{\text{path length (dm)} \times \text{concentration (g per 100 cm}^3)} = \frac{\alpha \times 100}{l \times c}$$

(where D = sodium D line i.e. light of $\lambda = 589$ nm; T = temperature in °C)

A 1:1 mixture of two enantiomers is known as a *racemate* or *racemic mixture* and this does not rotate plane-polarised light (it is optically inactive). The separation of a racemic mixture into its two enantiomers is called *resolution*.

For the fomation of a racemate in an S_N1 reaction, see Section 5.3.1.2

Example: 2-hydroxypropanoic acid (lactic acid)

(+)-enantiomer

$[\alpha]_D^{25} = +3.82$

(−)-enantiomer

$[\alpha]_D^{25} = -3.82$

The optical purity of a compound is a measure of the enantiomeric purity and this is described as the *enantiomeric excess (ee)*.

$$ee = \frac{\text{\% of the major}}{\text{enantiomer}} - \frac{\text{\% of the minor}}{\text{enantiomer}} = \frac{\text{observed } [\alpha]_D}{[\alpha]_D \text{ of pure enantiomer}} \times 100$$

For example, a 50% enantiomeric excess corresponds to a mixture of 75% of one enantiomer and 25% of the other enantiomer.

This is commonly called CIP nomenclature

3.3.2.2   The Cahn-Ingold-Prelog (R and S) nomenclature

The 3-dimensional arrangement of atoms attached to a stereogenic (or asymmetric) carbon centre is known as the *configuration*. The configuration can be assigned as *R* or *S* using the following *Cahn-Ingold-Prelog sequence rules*. This uses the same rules as for *E* and *Z* nomenclature (Section 3.3.1.2)

1. Rank the atoms directly attached to the stereogenic carbon atom in order of decreasing atomic number. The highest atomic number is ranked first.
2. If the atoms directly linked to the stereogenic centre are the same then the second, third, fourth, etc. atoms (away from the stereogenic centre) are ranked until a difference is found.
3. Multiple (double or triple) bonds are assumed to have the same number of single-bonded atoms.
4. The molecule is then orientated so that the group of lowest (fourth) priority is drawn pointing away from you. (Be careful not to swap the relative positions of the groups around.) The lowest priority group can now be ignored. A curved arrow is drawn from the highest priority group to the second and then to the third highest priority group. (If you have to draw a certain stereoisomer, always start with the lowest priority group pointing away from you.)

5. If the arrow is clockwise then the stereogenic centre has the *R* configuration. If the arrow is anticlockwise then the stereogenic centre has the *S* configuration. (The assignment of *R* may be remembered by analogy to a car steering wheel making a *right* turn.)

Examples (numbers on each carbon are given in italic to indicate priorities)

(R)-2-amino-2-hydroxymethylbutanoic acid

Naming carbon chains is discussed in Section 2.4

(S)-2-hydroxypropanoic acid or (S)-lactic acid

Hashed and wedged line notation is introduced in Section 3.3.2

It just so happens that (S)-lactic acid is the same as (+)-lactic acid (Section 3.3.2.1). However, the (+) or (−) sign of the optical rotation is not related to the R,S nomenclature (i.e. a (+)-enantiomer could be either R or S).

3.3.2.3 The D and L nomenclature

This old-fashioned nomenclature uses glyceraldehyde (or 2,3-dihydroxypropanal) as the standard. The (+)-enantiomer is given the label D (from *dextro*-rotatory), the (−)-enantiomer is given the label L (from *laevo*rotatory).

D-(+)-glyceraldehyde   L-(−)-glyceraldehyde

Notice that glyceraldehyde contains both primary and secondary alcohols together with an aldehyde (Section 2.1)

Any enantiomerically pure compound that is prepared from, for example, D-glyceraldehyde is given the label D. Any enantiomerically pure compound that can be converted into, for example, D-glyceraldehyde is given the label D.

D-Glyceraldehyde is the same as (R)-glyceraldehyde. However, the D or L label is not related to the R,S nomenclature (i.e. a D-enantiomer could be either R or S). This nomenclature is still used to assign natural products including sugars and amino acids (Sections 11.1 and 11.3).

3.3.2.4 Diastereoisomers (or diastereomers)

For compounds with two stereogenic centres, four stereoisomers are possible, as there are four possible combinations of R and S.

The R,S nomenclature is discussed in Section 3.3.2.2

R,R	R,S	S,R	S,S

- *Diastereoisomers* (or diastereomers) are stereoisomers that are not mirror images of each other. This means that diastereoisomers must have a different (R or S) configuration at *one* of the two stereogenic (chiral) centres.

Enantiomers often have different biological properties

- *Enantiomers* must have a different (*R* or *S*) configuration at *both* stereogenic (chiral) centres.

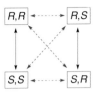

two enantiomer pairs linked by solid lines

four diastereoisomer pairs linked by dashed lines

Example (the numbers represent the numbering of the carbon backbone)

This compound is called 2,3-dihydroxy-4-methoxy-4-oxobutanoic acid

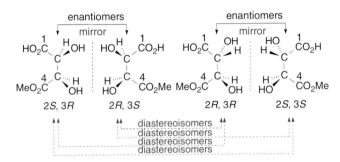

A compound with 3 chiral centres has 8 stereoisomers (4 diastereoisomers and 4 enantiomers)

For a compound containing *n* stereogenic (chiral) centres, the total number of stereoisomers will be 2^n and the number of pairs of enantiomers will be 2^{n-1}.

3.3.2.5 Diastereoisomers versus enantiomers

- Enantiomers have identical chemical and physical properties, except for their biological activity (i.e. they interact differently with other chiral molecules) and their effects on plane-polarised light.
- Diastereoisomers can have different physical (e.g. melting point, polarity) and chemical properties.
- Enantiomers are always chiral.

An achiral molecule has an identical mirror image

- Diastereoisomers can be chiral or achiral. If a diastereomer contains a plane of symmetry it will be achiral. These compounds, which contain stereogenic centres but are achiral, are called *meso compounds*. In effect, the plane of symmetry divides the molecule into halves, which contribute equally but oppositely to the rotation of plane-polarised light (i.e. they cancel each other out).

Bromination of (*E*)-but-2-ene forms a *meso*-1,2-dibromide (Section 6.2.2.2)

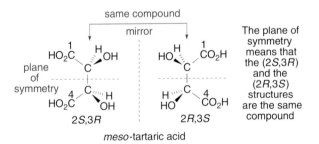

meso-tartaric acid

The plane of symmetry means that the (2*S*,3*R*) and the (2*R*,3*S*) structures are the same compound

3.3.2.6 Fischer projections

Molecules with asymmetric carbons can be represented by Fischer projections, in which a tetrahedral carbon atom is represented by two crossed lines. The vertical lines represent bonds pointing into the page (away from you) and the horizontal lines represent the bonds pointing out of the page (towards you).

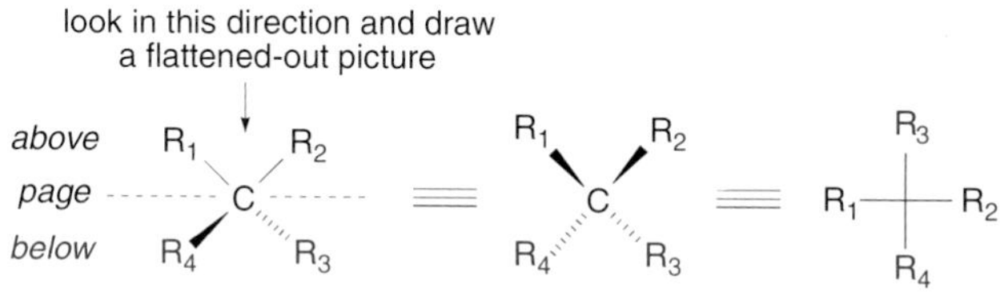

The compound is usually drawn so that the main carbon skeleton is vertical and the highest priority functional group is at the top of the vertical line.

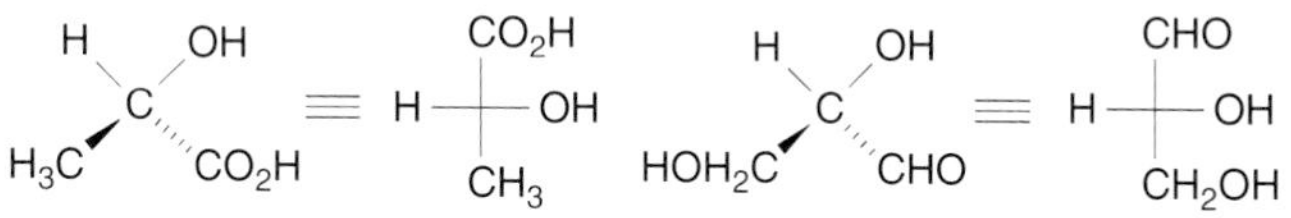

(R)-lactic acid

D-glyceraldehyde

The IUPAC name for lactic acid is 2-hydroxy-propanoic acid

This old-fashioned nomenclature is generally only used to show amino acids and sugars, which can contain several asymmetric centres (Sections 11.1 and 11.3).

Example: D-Erythrose or (2R,3R)-trihydroxybutanal

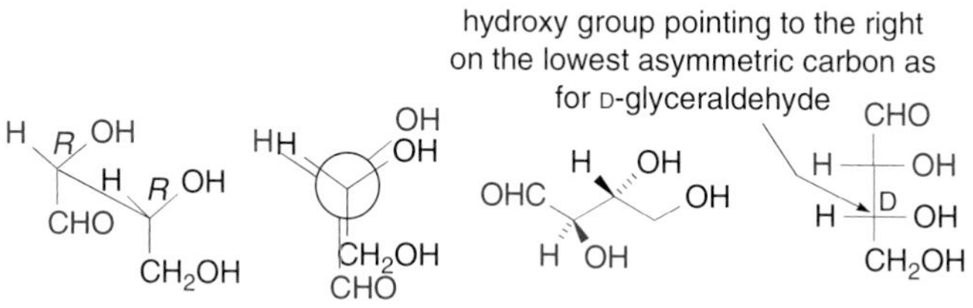

hydroxy group pointing to the right on the lowest asymmetric carbon as for D-glyceraldehyde

Sawhorse   Newman   Hashed-wedge   Fischer

D-Erythrose is an example of a carbohydrate (Section 11.1)

The *R,S* nomenclature can be assigned to Fischer projections by drawing the substituent of lowest priority in a vertical position (i.e. at the top). Although Fischer projections can be rotated in the plane of the paper by 180°, changing the position of substituents by 90° requires "double exchanges".

Examples (numbers on each carbon are given in italic to indicate priorities)

The IUPAC name for this compound is (*R*)-2-hydroxypropanal

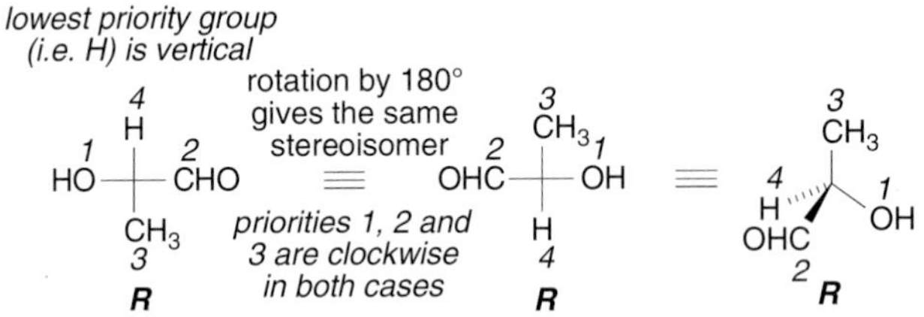

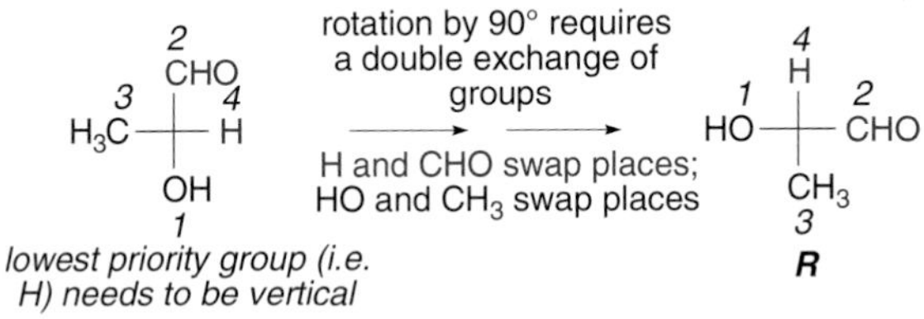

lowest priority group (i.e. H) is vertical

rotation by 180° gives the same stereoisomer

priorities 1, 2 and 3 are clockwise in both cases

rotation by 90° requires a double exchange of groups

H and CHO swap places; HO and CH₃ swap places

lowest priority group (i.e. H) needs to be vertical

Worked example

The following questions relate to compound **A** shown below.

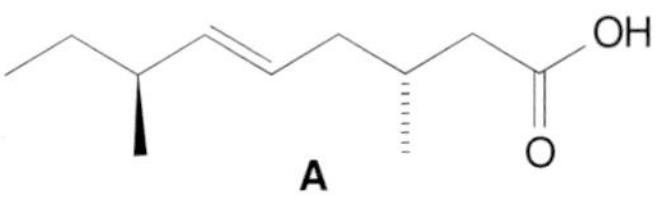

A

Hint: Assign priorities to the four substituents on the C=C bond

Hint: Assign priorities to the four groups attached to the chiral centre

Hint: Enantiomers have different configurations at both chiral centres; diastereoisomers have a different configuration at one of the two chiral centres

(a) Giving your reasons, assign *E* or *Z* configuration to the C=C bond in **A**.

(b) Giving your reasons, assign *R* or *S* configuration to each of the chiral centres in **A**.

(c) Draw the enantiomer of **A** and assign *R* or *S* configuration to each chiral centre.

(d) Draw a diastereoisomer of **A** and assign *R* or *S* configuration to each chiral centre.

Answer

(a)

Assignment of *E* and *Z* alkenes is discussed in Section 3.3.1.2

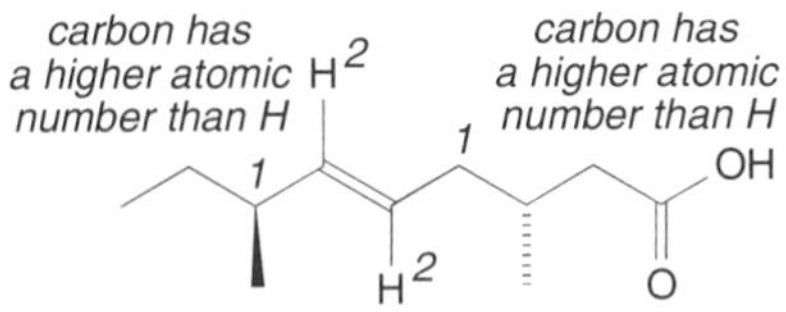

carbon has a higher atomic number than H

carbon has a higher atomic number than H

The groups of highest priority are on the opposite sides, therefore the C=C bond has *E*-configuration.

(b) There are two chiral centres in **A**.

group of lowest priority pointing away from you; the configuration is that indicated by the curved arrow

group of lowest priority pointing towards you; the configuration is the opposite of that indicated by the curved arrow

Assignment of R and S configuration is discussed in Section 3.3.2.2

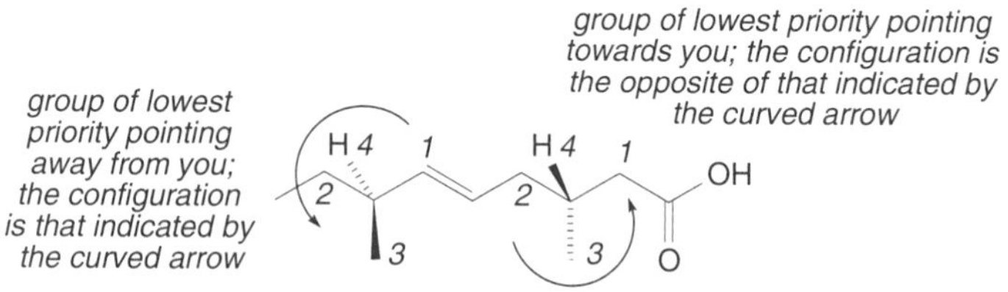

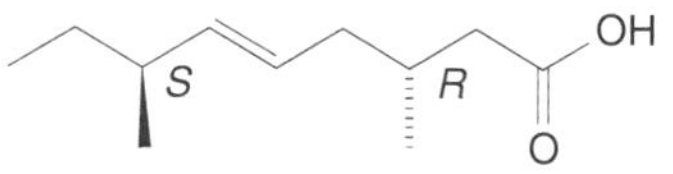

(c)

Enantiomers are introduced in Section 3.3.2.1

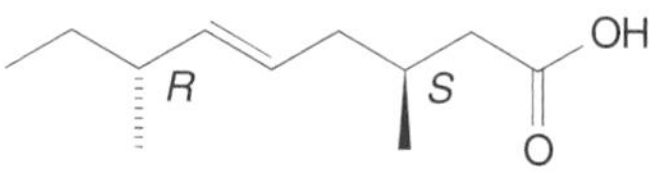

(d)

or

Diastereoisomers are introduced in Section 3.3.2.4

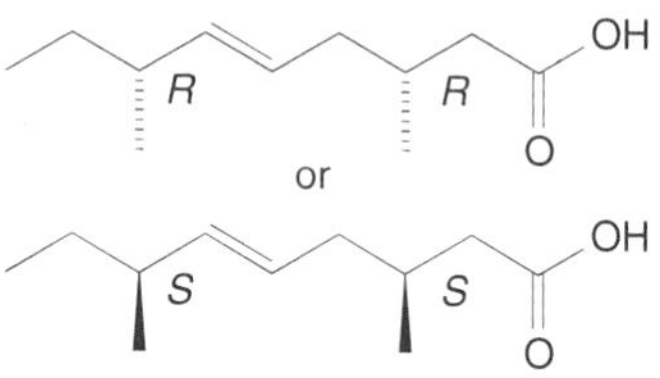

Problems

1. Consider the stereochemical priorities of the following groups using the Cahn-Ingold-Prelog system.

Section 3.3.1.2

$$-CONH_2, -CH_3, -CO_2H, -CH_2Br, -I, -CCl_3, -OCH_3$$

(a) Which group has the highest and which has the lowest priority?
(b) Which group ranks between $-CCl_3$ and $-CONH_2$?
(c) Assign prefixes Z or E as appropriate to each of compounds **A**, **B** and **C**.

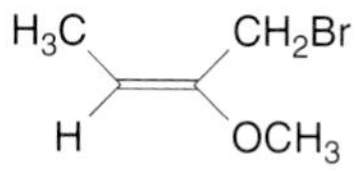

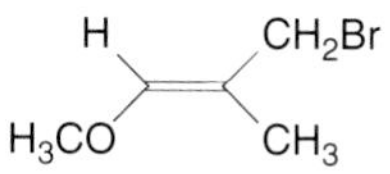

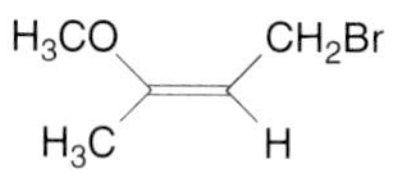

2. Assign the configuration R or S to compounds **D–F**, shown below.

Section 3.3.2.2

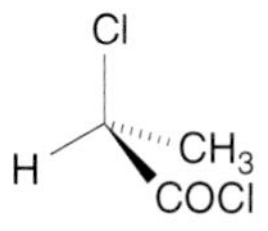

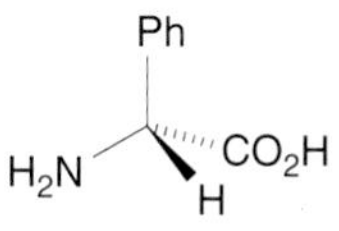

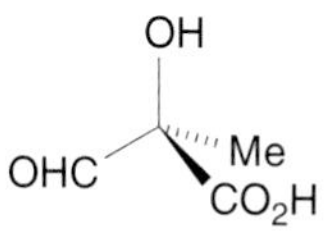

Section 3.3.2.5

Sections 3.2.1 and 3.2.2

3. The Fischer projection of *meso*-2,3-dihydroxybutane **G** is shown below.

(a) Why is this a *meso-* compound?

(b) Draw Newman and Sawhorse projections of the antiperiplanar conformation of **G**.

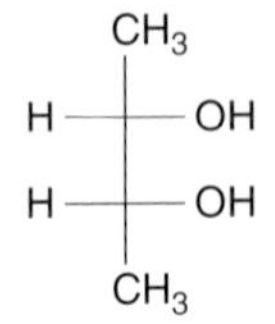

G

Section 3.3.2.6

Section 3.2.4

(c) Draw a Fischer projection of a diastereoisomer of **G**.

4. Draw the preferred chair conformation of (1*R*,2*S*,5*R*)-(−)-menthol (**H**).

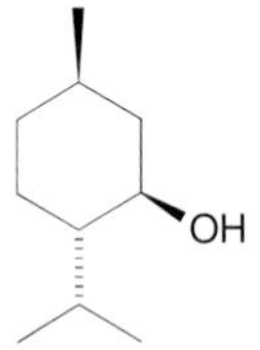

H

Section 3.3.2.4

5. What are the stereochemical relations (identical, enantiomers, diastereoisomers) of the following four molecules **I**–**L**? Assign absolute configurations at each stereogenic centre.

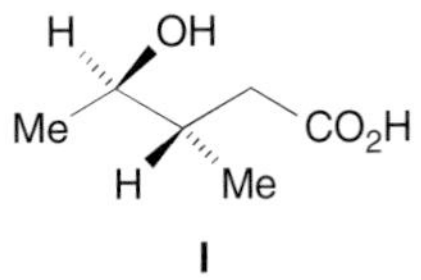

I

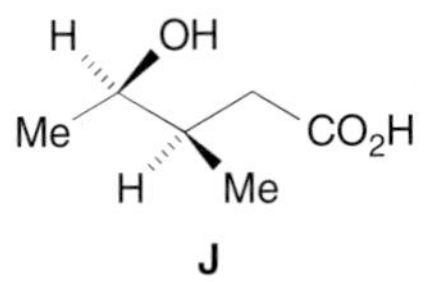

J

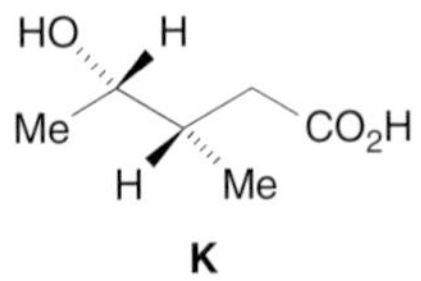

K

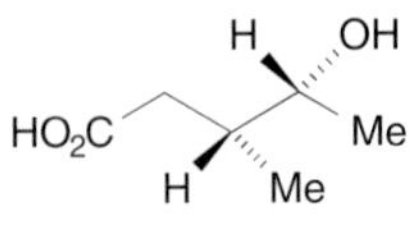

L

Sections 3.2.1 and 3.3.2.4

6. What are the stereochemical relations (identical, enantiomers, diastereoisomers) of the following four molecules **M**–**P**? Assign absolute
configurations at each stereogenic centre.

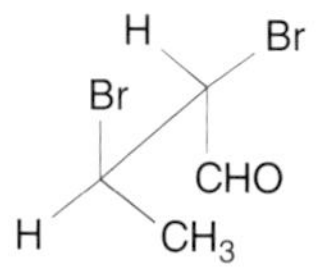

M

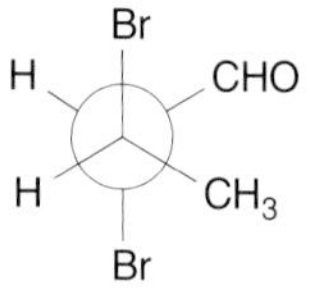

N

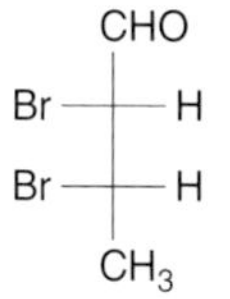

O

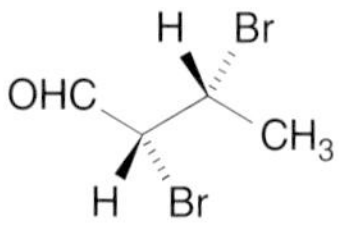

P

7. The following questions are based on the synthesis of the insect pheromone **T**, shown below.

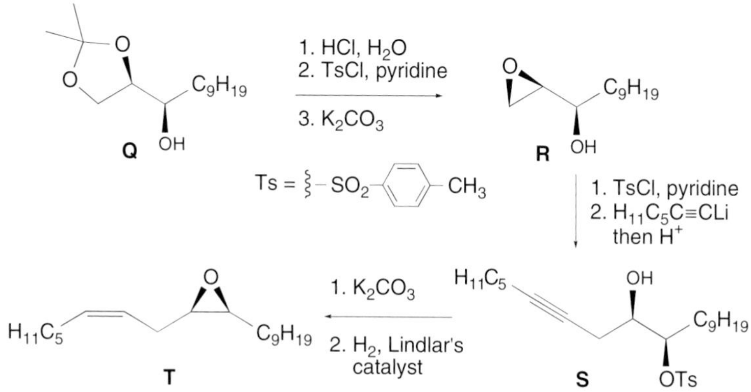

$Ts = \substack{\xi \\ \xi} - SO_2 -$⟨⟩$- CH_3$

(a) Assign *R* or *S* configuration to each of the chiral centres in compound **Q**.   Section 3.3.2.2
(b) Draw the enantiomer of compound **R**.   Section 3.3.2.1
(c) Draw a skeletal structure for $H_{11}C_5C\equiv C-Li$.   Section 3.2.2
(d) Assign *R* or *S* configuration to each of the chiral centres in compound **S**.   Section 3.3.2.2
(e) Draw the structure of a diastereoisomer of compound **S**.   Section 3.3.2.4
(f) Assign *E* or *Z* configuration to the C=C bond in compound **T**.   Section 3.3.1.2
(g) Assign *R* or *S* configuration to each of the chiral centres in compound **T**.   Section 3.3.2.2

4

Reactivity and mechanism

Key point. Organic reactions can take place by *radical* or, more commonly, by *ionic* mechanisms. The particular pathway of a reaction is influenced by the stability of the intermediate radicals or ions, which can be determined from an understanding of *electronic* and *steric* effects. For ionic reactions, *nucleophiles* (electron-rich molecules) form bonds to *electrophiles* (electron-poor molecules) and this can be represented using curly arrows. The energy changes that occur during a reaction can be described by the *equilibria* (i.e. how much of the reaction occurs) and also by the *rate* (i.e. how fast the reaction occurs). The position of the equilibrium is determined by the size of the *Gibbs free energy change* while the rate of a reaction is determined by the *activation energy.*

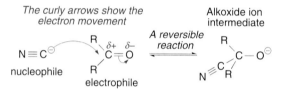

The curly arrows show the electron movement

Alkoxide ion intermediate

A reversible reaction

nucleophile

electrophile

The rate of addition of the nucleophile is influenced by the steric and electronic effects of the two R groups

4.1 Reactive intermediates: ions versus radicals

There are two ways of breaking a covalent bond. The unsymmetrical cleavage is called *heterolytic cleavage* (or heterolysis) and this leads to the formation of ions (positively charged cations and negatively charged anions). The symmetrical cleavage is called *homolytic cleavage* (or homolysis) and this leads to the formation of radicals.

Curly arrows can be used to represent bond cleavage. A double-headed arrow represents the movement of two electrons (and is used in polar reaction mechanisms). A single-headed arrow (or fish-hook) is used to represent the movement of

For an example of homolysis, see halogenation of alkanes (Section 5.2.1)

Keynotes in Organic Chemistry, Second Edition. Andrew F. Parsons.
© 2014 John Wiley & Sons, Ltd. Published 2014 by John Wiley & Sons, Ltd.

The stability of carbenium ions (R_3C^+), carbanions (R_3C^-) and carbon radicals ($R_3C^\bullet$) is discussed in Section 4.3

a single electron (and is used in radical reaction mechanisms). Curly arrows therefore always depict the movement of electrons.

Heterolysis (use one double-headed curly arrow)

$$A \frown B \longrightarrow A^\oplus + B^\ominus$$

cation   anion

Ions contain an even number of electrons

Both electrons in the two-electron bond move to only one atom

Homolysis (use two single-headed curly arrows)

$$A \frown\frown B \longrightarrow A^\bullet + B^\bullet$$

radical   radical

The two-electron bond is split evenly and one electron moves to each of the atoms

Radicals contain an odd number of electrons, and a dot ($\cdot$) is used to represent the unpaired electron.

Processes that involve unsymmetrical bond cleavage or bond formation are known as *polar* (or *ionic*) *reactions*. Processes that involve symmetrical bond cleavage or bond formation are known as *radical reactions*.

For an example of heterolytic bond formation, see the S_N1 reaction (Section 5.3.1.2)

Heterolytic bond formation (the reverse of heterolysis)

$$A^\oplus \quad B^\ominus \longrightarrow A - B$$

cation   anion

The arrow points to the middle of the bond that is formed, or directly to the cation

A two-electron arrow goes from the anion/lone pair to the cation. The arrow points from the electron-rich centre to the electron-poor centre

Homolytic bond formation (the reverse of homolysis)

The reaction of two radicals is called coupling (Section 4.6.2)

$$A^\bullet \quad B^\bullet \longrightarrow A - B$$

radical   radical

The arrows point to the middle of the bond that is formed

The one-electron arrows point towards one another to make a new two-electron bond. Each arrow points away from the dot and finishes midway between A and B

Formally, the two-electron arrow should point to where the new bond will be formed if the electrons are being used to form the bond. If the electrons end up as a lone pair then they point to the atom. Many textbooks, including this one, show double-headed arrows drawn directly on to the atom in both cases.

When more than one two-electron arrow is used in a reaction scheme, the arrows *must* always point in the same direction.

This reaction involves attack of a nucleophile to open the epoxide ring (Section 6.2.2.6)

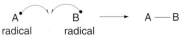

For any reaction, the overall charge of the starting materials should be the *same* as the product(s).

4.2 Nucleophiles and electrophiles

- *Nucleophiles* are electron-rich species that can form a covalent bond by *donating* two electrons to an electron-poor site. Nucleophiles are negatively charged (anions) or neutral molecules that contain a lone pair of electrons.

- *Electrophiles* are electron-poor species that can form a covalent bond by *accepting* two electrons from an electron-rich site. Electrophiles are often positively charged (cations) although they can also be neutral.

Sometimes, negatively charged nucleophiles are drawn showing both a negative charge and two dots (to show the lone pair)

Nucleophile + Electrophile   Nucleophile + Electrophile
(electron-rich) (electron-poor)   (electron-rich) (electron-poor)

The nucleophilic or electrophilic sites within a neutral organic molecule can be determined by: (i) the presence of lone pairs of electrons; (ii) the type of bonding (sp, sp^2 or sp^3); and/or (iii) the polarity of the bonds.

sp, sp^2 and sp^3 hybridisation is introduced in Section 1.5

1. An atom (such as nitrogen, oxygen or sulfur) bearing an electron pair will be a nucleophilic site.
2. Double or triple carbon–carbon bonds in alkenes and alkynes, as well as aromatic rings, are of high electron density and so are nucleophilic sites. (Single C–C bonds in alkanes are not nucleophilic.)
3. In a polar bond (Section 1.6.1), the electrons are held closer to the more electronegative atom. The electronegative atom will be a nucleophilic site, and the less electronegative atom will be an electrophilic site.

Electrophiles react with the electron-rich C=C bond in an alkene (Section 6.2.2)

Single bonds

Y is *more* electronegative than carbon	Y is *less* electronegative than carbon
$\delta+$ $\delta-$ C—Y Y = Cl, Br, O, N	$\delta-$ $\delta+$ C—Y Y = Mg, Li
electrophilic site	**nucleophilic site**

Double and triple bonds

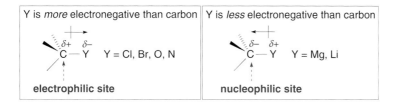

$\delta+$ $\delta-$ C=Y   Y = O, NH, NR

electrophilic site

$\delta+$ $\delta-$ C≡N

electrophilic site

The arrows above the bonds indicate the polarisation of electrons – the arrow points towards the more electron-rich part (Section 1.6.1)

4.2.1　Relative strength

4.2.1.1　Nucleophiles

Conjugate acids are introduced in Section 1.7

A nucleophile with a negative charge is always a more powerful nucleophile than its conjugate acid (which is neutral).

$$\text{Nucleophilic strength} \quad HO^{\ominus} > H_2\ddot{O}$$

The relative nucleophilic strength (or nucleophilicity) of an anion, or a nucleophilic site within a neutral molecule, depends on the availability of the pair of electrons. The more electronegative the atom, the less nucleophilic the atom will be, because the electrons are held tighter to the nucleus.

The nucleophilic strength of anions, within the same row of the periodic table, follows the same order as basicity: the more electronegative the atom bearing the negative charge, the weaker the nucleophile, and the weaker the base.

$$\text{Nucleophilic strength} \quad R_3C^{\ominus} > R_2N^{\ominus} > RO^{\ominus} > F^{\ominus}$$
$$\text{(or basicity)}$$

$$\text{Electronegativity} \quad C < N < O < F$$

For the Pauling electronegativity scale, see Section 1.6.1

$$\text{Nucleophilic strength} \quad R-\ddot{N}H_2 > R-\ddot{O}H > R-\ddot{F}$$
$$\text{(or basicity)}$$

Basicity = donation of a pair of electrons to H (or H⁺)	**Nucleophilicity** = donation of a pair of electrons to an atom other than H

For a table of pK_a values, see Appendix 3

The pK_a values can therefore be used to estimate the nucleophilicity of atoms within the same row. It is not exact because the nucleophilicity is strongly affected by steric factors (see Section 4.4), while the basicity is not.

The nucleophilic strength of anions and neutral atoms increases on going down a group of the periodic table. The electrons are held less tightly to the nucleus as the atom size increases, and hence they are more available for forming bonds. Larger atoms, with more loosely held electrons (than smaller atoms), are said to have a higher *polarisability*.

The iodide ion is a good nucleophile (Section 5.3.1.5)

$$\text{Nucleophilic strength} \quad \begin{array}{c} H_2\ddot{S} > H_2\ddot{O} \\ I^{\ominus} > Br^{\ominus} > Cl^{\ominus} > F^{\ominus} \end{array}$$

Anions are very weakly solvated in Me₂S=O, because the δ+ sulfur is in the middle of the molecule and anions cannot get close to it

The nucleophilic strength of anions depends on the solvent.

- Anions are generally more nucleophilic in *aprotic solvents* (these contain polar groups but no O–H or N–H bonds), such as dimethyl sulfoxide (Me₂SO), than in protic solvents.
- In *protic solvents* (these contain polar groups and O–H or N–H bonds), such as methanol (MeOH), the solvent can form hydrogen bonds to the anion.

This lowers the nucleophilicity because a solvent shell surrounds the anion and this hinders attack at the electrophile. Large anions are less solvated, and so are stronger nucleophiles than small anions in protic solvents. For example, the smaller F^- ion is more heavily solvated than I^- and so is a weaker nucleophile.

Solvation is the interaction of a solute with the solvent

4.2.1.2 Electrophiles

An electrophile with a positive charge is always a more powerful electrophile than its conjugate base (which is neutral).

The double-headed straight arrow indicates that the two structures are resonance forms (Section 1.6.3)

Electrophilic strength

protonated ketone   ketone

The relative electrophilic strength (or electrophilicity) of a cation depends on the stability of the positive charge. Inductive (+I), mesomeric (+M) and/or steric effects (see Section 4.4) can all lower the reactivity of the cation.

Inductive and mesomeric effects are introduced in Sections 1.6.1 and 1.6.3

The relative electrophilic strength of an electrophilic site within a neutral molecule depends on the size of the partial positive charge ($\delta+$). Carbon atoms are electrophilic when attached to electronegative atoms ($-I$ groups). The more electronegative the atom(s) bonded to carbon, the more electrophilic the carbon atom.

Electrophilic strength

acyl chloride   ketone

The (carbonyl) carbon atom of the acyl (acid) chloride is more electrophilic because both oxygen and chlorine atoms are electronegative

The relative reactivity of carbonyl compounds is discussed further in Sections 8.3.1 and 9.3.1

4.3 Carbocations, carbanions and carbon radicals

Carbocations, which include *carbenium* and *carbonium* ions, contain a positive charge on carbon. Carbenium ions have three bonds to the positively charged carbon (e.g. Me_3C^+), while carbonium ions contain five bonds (e.g. H_5C^+). (Carbenium ions are the most important.)

- *Carbenium ions* (R_3C^+) are generally planar and contain an empty p orbital. They are stabilised by electron-donating groups (R is +I and/or +M), which delocalise the positive charge; +M groups are generally more effective than +I groups.

Stabilisation of carbenium ions by +M effects is introduced in Section 1.6.3

planar (sp²)   planar (sp²)   pyramidal (sp³)

carbocation   carbanion

The formation and stability of carbanions, called enolate ions, from carbonyl compounds is discussed in Section 8.4.3

- *Carbanions* have three bonds on the carbon atom, which bears the negative charge. Carbanions (R_3C^-) can be planar (sp^2) or pyramidal (sp^3) (or something in-between). They are stabilised by: (i) electron-withdrawing groups (R is –I and/or –M); (ii) an increase in "s" character of the carbon bearing the negative charge; (iii) aromatisation, which delocalises the negative charge (see Section 1.7).
- *Carbon radicals* have three bonds on the carbon atom, which contains the unpaired electron. Carbon radicals ($R_3C^\bullet$) are generally planar (sp^2). Like carbenium ions, they are stabilised by electron-donating groups (R is +I); they are also stabilised by unsaturated groups (e.g. R is Ph or $COCH_3$) that delocalise the unpaired electron.

Resonance forms are introduced in Section 1.6.3

Carbon radicals are similar to carbenium ions because both contain an electron deficient carbon atom. For carbenium ions the carbon atom is deficient of two electrons, for carbon radicals the carbon atom is deficient of one electron.

4.3.1 Order of stability

Alkyl groups (R) include methyl, ethyl, propyl and butyl (Section 2.2)

The order of stability of carbocations, carbanions and carbon radicals bearing electron donating (+I) alkyl groups, R, is as follows.

Stabilisation of carbanions by resonance is introduced in Section 1.6.3

Carbanions can be stabilised by electron-withdrawing groups (–I, –M groups), whereas carbocations can be stabilised by electron-donating groups (+I, +M groups).

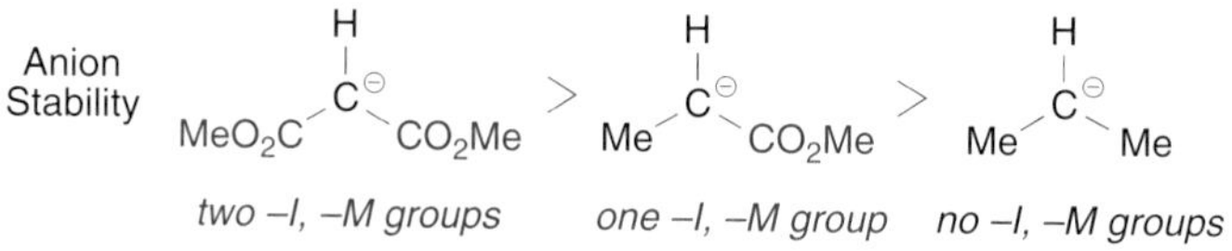

Anion
Stability

two –I, –M groups   *one –I, –M group*   *no –I, –M groups*

For the formation and reaction of
ester enolate ions see Section 9.11

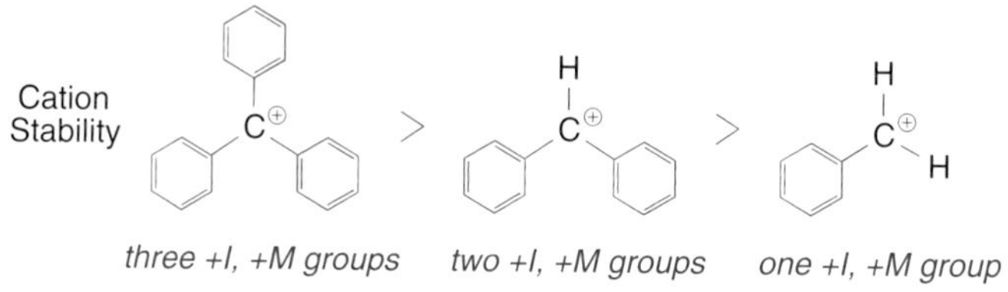

Cation
Stability

three +I, +M groups   *two +I, +M groups*   *one +I, +M group*

Ph_3C^+ is commonly called the
trityl cation, whereas $PhCH_2^+$ is
the benzyl cation (Section 5.3.1.2)

4.4 Steric effects

The size as well as the electronic properties (i.e. inductive and mesomeric effects)
of the surrounding groups affects the stability of carbocations, carbanions and
radicals. When bulky substituents surround a cation, for example, this reduces the
reactivity of the cation to nucleophilic attack by *steric effects*. This is because the
bulky groups hinder the approach of a nucleophile.

For steric strain in cyclohexanes,
see Section 3.2.4

When the size of groups is responsible for reducing the reactivity at a site
within a molecule this is attributed to *steric hindrance*. When the size of groups is
responsible for increasing the reactivity at a site within a molecule this is
attributed to *steric acceleration*.

Steric effects explain why tertiary
halogenoalkanes undergo S_N1
reactions (Section 5.3.1.1)

Electronic and/or steric effects can explain the particular pathway for any
given reaction.

4.5 Oxidation levels

- An *organic oxidation reaction* involves either (i) a decrease in the hydrogen
  content or (ii) an increase in the oxygen, nitrogen or halogen content of a
  molecule. Electrons are lost in oxidation reactions.
- An *organic reduction reaction* involves either (i) an increase in the hydrogen
  content or (ii) a decrease in the oxygen, nitrogen or halogen content of a
  molecule. Electrons are gained in reductions.

We can classify different functional groups by the *oxidation level* of the
carbon atom within the functional group, using the following guidelines.

Functional groups are introduced in
Section 2.1

1. The greater the number of heteroatoms (e.g. O, N, halogen) attached to the
   carbon, the higher the oxidation level. Each bond to a heteroatom increases the
   oxidation level by $+1$.
2. The higher the degree of multiple-bonding the higher the oxidation level of the
   carbon.

Oxidation level	0	1	2	3	4
Functional Groups	R_4C alkane	$RHC = CHR$ alkene	$RC \equiv CR$ alkyne		
		RCH_2X monohalide	$RCHX_2$ dihalide	RCX_3 trihalide	CX_4 tetrahalide
		RCH_2OR ether	$RCH(OR)_2$ acetal		
		RCH_2OH primary alcohol	$\overset{\displaystyle O}{\overset{\|}{R-C-R}}$ ketone	$\overset{\displaystyle O}{\overset{\|}{R-C-OH}}$ carboxylic acid	
		RCH_2NH_2 primary amine	$RCH = NH$ imine	$RC \equiv N$ nitrile	

RCH$_2$X is a primary halogenoalkane, where is F, Cl, Br, or I (Section 2.1)

- Carbon atoms (within functional groups) at the same oxidation level can be interconverted *without* oxidation or reduction.

Example (numbers on each carbon are given in italic to indicate oxidation levels)

Hydrolysis of nitriles (RCN) to form carboxylic acids (RCO$_2$H) is discussed in Section 9.9

$$RC \equiv N \longrightarrow \overset{\displaystyle O}{\overset{\|}{R-C-OH}} \longrightarrow \overset{\displaystyle O}{\overset{\|}{R-C-OR}}$$
$$\quad 3 \qquad\qquad 3 \qquad\qquad 3$$

No oxidation nor reduction required

- Carbon atoms (within functional groups) at different oxidation levels are interconverted *by* oxidation (to increase the oxidation level) or reduction (to decrease the oxidation level).

Example (numbers on each carbon are given in italic to indicate oxidation levels)

Reduction of an alkene to form an alkane, using H$_2$/Pd/C, is discussed in Section 6.2.2.10

$\longrightarrow$ *oxidation*

$$RH_2C - CH_2R \longrightarrow RHC = CHR \longrightarrow RC \equiv CR$$
$$\quad 0 \quad\; 0 \qquad\qquad 1 \quad\; 1 \qquad\qquad 2 \quad\; 2$$

$$RCH_2OH \longrightarrow RCHO \longrightarrow RCO_2H$$
$$\quad 1 \qquad\qquad 2 \qquad\qquad 3$$

$\longleftarrow$ *reduction*

4.6 General types of reaction

4.6.1 Polar reactions (involving ionic intermediates)

4.6.1.1 Addition reactions

Addition reactions occur when two starting materials add together to form only one product.

$$A \quad + \quad B \longrightarrow A\text{–}B$$

The mechanism of these reactions can involve an initial electrophilic or nucleophilic attack on to the key functional group.

Electrophilic addition to alkenes (Section 6.2.2).

Nucleophilic addition to aldehydes and ketones (Section 8.3).

$R_2C(OH)_2$ is called a 1,1-diol or a hydrate (Section 8.3.5.1)

4.6.1.2 Elimination reactions

Elimination reactions are the opposite of addition reactions. One starting material is converted into two products.

$$A–B \longrightarrow A + B$$

The mechanism of these reactions can involve loss of a cation or anion to form ionic reaction intermediates.

Elimination of halogenoalkanes (Section 5.3.2).

4.6.1.3 Substitution reactions

Substitution reactions occur when two starting materials exchange groups to form two new products.

$$A — B + C — D \longrightarrow A — C + B — D$$

The mechanism of these reactions can involve an initial electrophilic or nucleophilic attack on to the key functional group.

Nucleophilic substitution of halogenoalkanes (Section 5.3.1).

This is an example of an S_N2 reaction (Section 5.3.1.1)

Electrophilic substitution of benzene (Section 7.2).

Lewis acids are defined in
Section 1.7.3

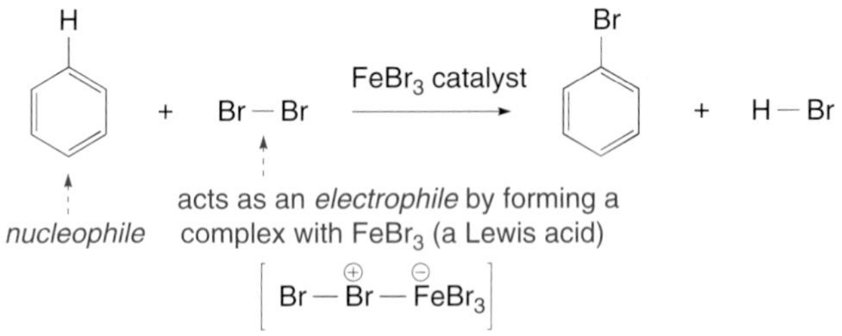

acts as an *electrophile* by forming a
nucleophile   complex with FeBr₃ (a Lewis acid)

4.6.1.4 Rearrangement reactions

Isomers are introduced in
Section 3.1

Rearrangement reactions occur when one starting material forms one product
with a different arrangement of atoms and bonds (i.e. the product is an isomer of
the starting material).

$$ A \longrightarrow B $$

The stability of carbocations is
discussed in Section 4.3.1

The mechanism of these reactions often involves carbocation intermediates
and the first-formed cation (e.g. primary or secondary) can rearrange to a more
stable cation (e.g. tertiary).

Me₃C⁺ is called the *tert*-butyl
cation

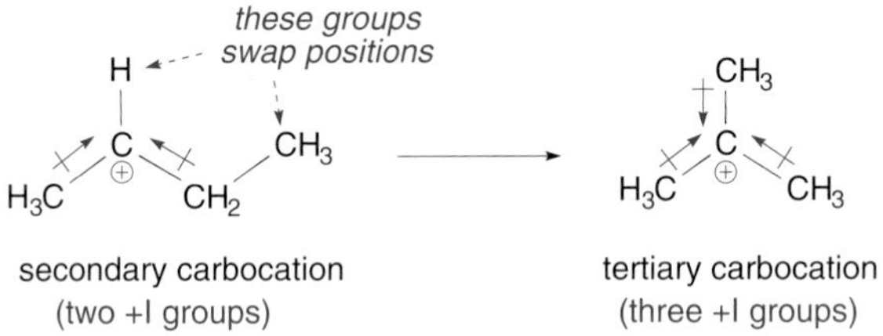

secondary carbocation
(two +I groups)

tertiary carbocation
(three +I groups)

4.6.2 Radical reactions

Radicals, like ions, can undergo addition, elimination, substitution and rearrangement reactions. A radical reaction comprises a number of steps.

1. *Initiation* – the formation of radicals by homolytic bond cleavage (this generally
   requires heat or light).

Homolytic bond cleavage
(homolysis) is introduced in
Section 4.1

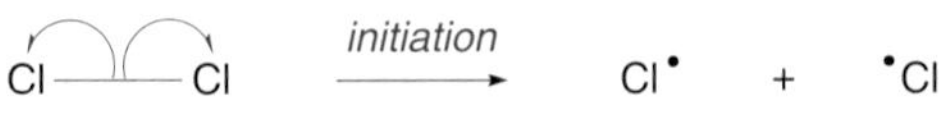

2. *Propagation* – the reaction of a radical to produce a new product radical. This
   may involve an addition, elimination, substitution or rearrangement reaction.

propagation reactions

Radical reactions involve breaking
weaker bonds to form stronger
bonds

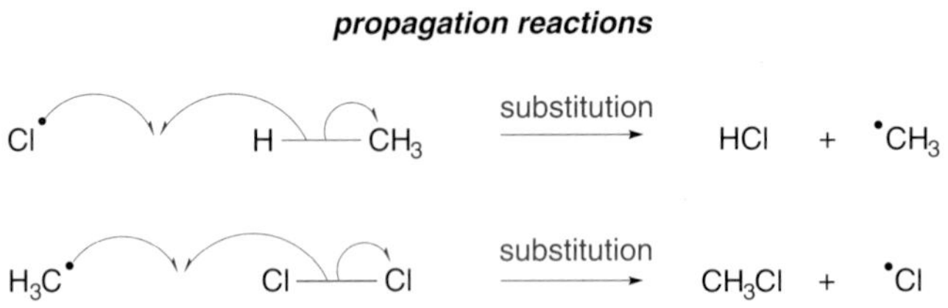

3. *Termination* – the coupling of two radicals to form only non-radical products.

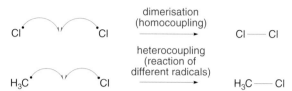

The chlorination of alkanes, such as methane (CH_4), leads to the substitution of hydrogen atoms for chlorine atoms in a radical chain reaction (Section 5.2.1).

4.6.3 Pericyclic reactions

Pericyclic reactions take place in a single step without (ionic or radical) intermediates and involve a cyclic redistribution of bonding electrons.

The Diels-Alder cycloaddition reaction (Section 6.2.2.11).

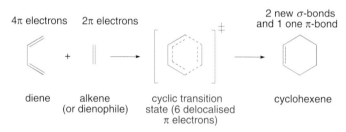

The Diels-Alder reaction is an example of a 4+2 cycloaddition: one reactant with 4π electrons reacts with another reactant with 2π electrons

4.7 Ions versus radicals

- *Heterolytic cleavage* of bonds occurs at room temperature in polar solvents. The ions which are formed are solvated (i.e. a solvent shell surrounds them) and stabilised by polar solvents.
- *Homolytic cleavage* of bonds occurs at high temperature in the absence of polar solvents. When a compound is heated in a non-polar solvent, radicals are formed. Radicals are uncharged and so have little interaction with the solvent. The energy required to cleave a bond homolytically, to give radicals, is called the *bond dissociation enthalpy (energy)* or *bond strength*. The lower the bond dissociation enthalpy, the more stable the radicals (as they are easier to form).
- *Polar (ionic) reactions* occur because of electrostatic attraction; a positive or δ+ charge attracts a negative or δ– charge. Electron-rich sites react with electron-poor sites.
- *Radical reactions* occur because radicals, which have an odd number of electrons in the outer shell, need to 'pair' the electron to produce a filled outer shell.

Heterolytic bond cleavage and homolytic bond cleavage is introduced in Section 4.1

4.8 Reaction selectivity

- *Chemoselectivity* – reaction at one functional group in preference to another functional group(s).

The oxidation level of the C=O carbon is 2, whereas the C–OH carbon is 1 (Section 4.5)

Example: reduction of a ketoester (Section 8.3.3.1)

selective reduction of the ketone to form the secondary alcohol

Isomers are introduced in Section 3.1

- *Regioselectivity* – reaction at one position within a molecule in preference to others. This leads to the selective formation of one structural isomer (regioisomer).

Example: addition of HCl to an unsymmetrical alkene (Section 6.2.2.1)

Me stands for methyl, or CH_3

regioselective addition of the H atom onto the carbon at the end of the double bond and the Cl atom onto the carbon with the most alkyl substituents

- *Stereoselectivity* – the formation of one enantiomer, one diastereoisomer or one double bond isomer in preference to others.

Et stands for ethyl, or CH_3CH_2

Example: catalytic hydrogenation of an alkyne (Section 6.3.2.4)

E and *Z* alkenes are defined in Section 3.3.1.2

formation of a *Z*-alkene rather than a *E*-alkene

S_N2 reactions are stereospecific (Section 5.3.1.1)

In a stereospecific reaction, different stereoisomers react differently.

4.9 Reaction thermodynamics and kinetics

The *thermodynamics* of a reaction tells us in what direction the reaction proceeds (and how much energy will be consumed or released). The *kinetics* of a reaction tells us whether the reaction is fast or slow.

4.9.1 Thermodynamics

4.9.1.1 Equilibria

All chemical reactions can be written as equilibrium processes, in which forward and backward reactions occur concurrently to give an equilibrium position.

Drawing arrows pointing in opposite directions denotes the equilibrium and the position of the equilibrium is expressed by the *equilibrium constant (K)*. For simplicity, this is defined below in terms of concentration rather than activity (as activity is approximately equal to concentration in dilute solution).

Many organic reactions are reversible. This includes the formation of acetals (Section 8.3.5.2) and esters (Section 9.4.2); the equilibrium can be driven to reactants or products by changing the reaction conditions

$$A \rightleftharpoons B \quad K = \frac{\text{concentration of product(s) at equilibrium}}{\text{concentration of reactant(s) at equilibrium}} = \frac{[B]_{eq}}{[A]_{eq}}$$

If K is larger than 1, then the concentration of B will be larger than the concentration of A at equilibrium.

The size of K is related to the free energy difference between the starting materials and products. For the efficient conversion of A into B, a high value of K is required. This means that the free energy of the products (in kJ mol^{-1}) must be lower than the free energy of the starting materials. The total free energy change (under standard conditions) during a reaction is called the *standard Gibbs free-energy change, $\Delta_r G^\theta$ (in kJ mol^{-1})*.

$\Delta_r G^\theta$ means the Gibbs free-energy change during the reaction under standard conditions of 1 bar

$$\Delta_r G^\ominus = \text{free energy of products} - \text{free energy of reactants}$$

$$\Delta_r G^\ominus = -RT \ln K \quad \begin{array}{l} R = \text{gas constant } (8.314 \text{ J K}^{-1}\text{mol}^{-1}) \\ T = \text{absolute temperature (in K)} \end{array}$$

The standard free energy change for a reaction is a reliable guide to the extent of a reaction, provided that equilibrium is reached.

- If $\Delta_r G^\theta$ is *negative*, then the *products* will be favoured at equilibrium ($K > 1$).
- If $\Delta_r G^\theta$ is *positive*, then the *reactants* will be favoured at equilibrium ($K < 1$).
- If $\Delta_r G^\theta$ is *zero*, then $K = 1$, hence there will be the same concentration of reactants and products.

$\Delta_r G^\theta$ is sometimes referred to as the 'driving force' of a chemical reaction

At a particular temperature, K is constant.

Remember that K and $\Delta_r G^\theta$ say nothing about the rate of a reaction. The sign and magnitude of $\Delta_r G^\theta$ decides whether an equilibrium lies in one direction or another.

4.9.1.2 Enthalpy and entropy

The standard Gibbs free-energy change of reaction, $\Delta_r G^\theta$, is related to the *enthalpy change of reaction ($\Delta_r H^\theta$)* and *the entropy change of reaction ($\Delta_r S^\theta$)*. The change in free energy for a reaction at a given temperature has contributions from *both* the change in enthalpy and the change in entropy.

Notice that $\Delta_r H^\theta$ and $\Delta_r S^\theta$ are independent of one another; if they have the same sign, they will work against one another

$$\Delta_r G^\ominus = \Delta_r H^\ominus - T\Delta_r S^\ominus \quad T = \text{absolute temperature (in K)}$$

Enthalpy

The enthalpy change of reaction is the heat exchanged with the surroundings (at constant pressure) in a chemical reaction. This represents the difference in stability (bond strength) of the reagents and products.

For example, chlorination of methane (CH_4) using Cl_2 and UV radiation is exothermic (Section 5.2.1); the bonds formed (C-Cl and H-Cl) are stronger than those broken (C-H and Cl-Cl)

- If $\Delta_r H^\theta$ is negative, then the bonds in the product(s) are *stronger* overall than those in the starting material. Heat is released in an *exothermic* reaction.
- If $\Delta_r H^\theta$ is positive then the bonds in the product(s) are *weaker* overall than those in the starting material. Heat is absorbed in an *endothermic* reaction.

Entropy

The entropy change of reaction provides a measure of the change in molecular disorder or randomness caused by a reaction.

- $\Delta_r S^\theta$ is negative when the reaction leads to *less* disorder. This occurs when two reactants are converted into one product.
- $\Delta_r S^\theta$ is positive when the reaction leads to *more* disorder. This occurs when one reactant is converted into two products.

Gibbs free energy

For a negative value of $\Delta_r G^\theta$ (in which products are favoured over reactants at equilibrium) we require low positive, or high negative, values of $\Delta_r H^\theta$ and high positive values of $T\Delta_r S^\theta$.

- If the reaction is *exothermic* then $\Delta_r H^\theta$ will be *negative*, and hence $\Delta_r G^\theta$ will be negative, provided $\Delta_r S^\theta$ is not large and negative.
- If the reaction is *endothermic* then $\Delta_r H^\theta$ will be *positive* and the $-T\Delta_r S^\theta$ term will need to be larger than this for $\Delta_r G^\theta$ to be negative. This will require a large, positive entropy change ($\Delta_r S^\theta$) and/or a high reaction temperature (T).

4.9.2 Kinetics

4.9.2.1 Reaction rate

See, for example, the transition state for an S_N2 reaction in Section 5.3.1.1 and the transition state for an E2 reaction in Section 5.3.2.1

Although a negative value of $\Delta_r G^\theta$ is required for a reaction to occur, the rate at which it occurs is determined, from transition state theory, by the *Gibbs energy of activation* $\Delta^\ddagger G$. This is the energy difference between the reactants and the *transition state*. A transition state is a structure that represents an energy maximum on converting reactants into products and it cannot be isolated. It is often drawn in square brackets with a double-dagger superscript ($\ddagger$). (This is not the same as a reaction *intermediate*, which occurs at a local energy minimum and can be detected and sometimes isolated (see below)).

We can show the energy changes that occur during a reaction by a Gibbs energy profile.

$\Delta^\ddagger G^\theta$ is the standard Gibbs energy of activation

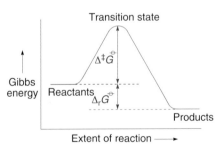

The higher the Gibbs energy of activation energy the slower the reaction. Reactions having activation energies above $20\,kJ\,mol^{-1}$ generally require heating so reactants can be converted into products. An increase in temperature therefore increases the rate of the reaction. Most organic reactions have activation energies of between 40–$150\,kJ\,mol^{-1}$.

The activation energy E_a (in $J\,mol^{-1}$) can be determined experimentally by measuring the *rate constant* of the reaction at different temperatures using the *Arrhenius equation*.

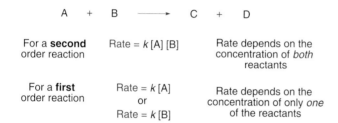

$$k = Ae^{-E_a/RT}$$

k = rate constant R = gas constant ($8.314\ J\ K^{-1}\,mol^{-1}$)
A and e = constants T = temperature (K)

- The rate constant for the reaction, k, can be determined by monitoring the rate at which reactants disappear/products appear (at constant temperature) by varying the concentration of the reactant or reactants.

$$A\ +\ B\ \longrightarrow\ C\ +\ D$$

For a **second** order reaction	Rate = k [A] [B]	Rate depends on the concentration of *both* reactants
For a **first** order reaction	Rate = k [A] or Rate = k [B]	Rate depends on the concentration of only *one* of the reactants

An S_N1 reaction is first order (Section 5.3.1.2); the reaction mechanism involves two steps, and the formation of an intermediate. The first step is the rate-determining step.

Organic reactions generally consist of a number of successive steps. The slowest step (which leads to the highest energy transition state) is called the *rate-determining step* and this is the reaction rate that can be determined experimentally.

Organic reactions with a number of steps will have *intermediates*. These represent a localised minimum energy in the reaction profile. An energy barrier must be overcome before the intermediate forms a more stable product(s) or a second intermediate.

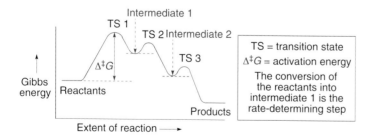

TS = transition state
$\Delta^{\ddagger}G$ = activation energy
The conversion of the reactants into intermediate 1 is the rate-determining step

Notice that this reaction profile shows the products at a lower energy that the reactants and that only TS1 (the highest energy transition state) matters for determining the reaction rate

The structure of the intermediates can give us an idea of the structure of the transition states. The *Hammond postulate* says that the structure of a transition state resembles the structure of the nearest stable species.

- For an *exothermic* reaction (or step of a reaction), the transition state resembles the structure of the *reactant*. This is because the energy level of the transition state is closer to the reactant(s), than to the product(s).

Radical halogenation of alkanes, which involves radical intermediates, is discussed in Section 5.2.1

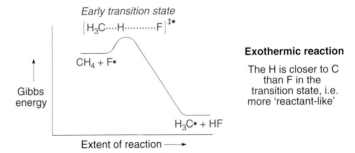

- For an *endothermic* reaction (or step of a reaction), the transition state resembles the structure of the *product*. This is because the energy level of the transition-state is closer to the product than to the reactant.

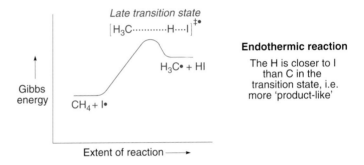

In reactions of carbonyls, an acid catalyst is often used; protonation of the C=O bond forms a stronger electrophile (see, for example, Section 8.3.5)

Catalysts increase the rate of a reaction by allowing the reaction to proceed by a different pathway, which has a lower energy transition state. Although the catalyst affects the rate at which equilibrium is established, it does not alter the position of equilibrium. The addition of a catalyst, which is not chemically changed during the reaction, allows many reactions to take place more quickly and at lower temperatures.

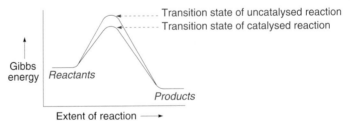

A heterogeneous catalyst is used in the hydrogenation of alkenes (Section 6.2.2.10)

A *homogeneous* catalyst is in the *same* phase as the starting materials of the reaction that it is catalysing.

A *heterogeneous* catalyst is in a *different* phase from the starting materials of the reaction that it is catalysing.

4.9.3 Kinetic versus thermodynamic control

For a reaction that can give rise to more than one product, the amount of each of the different products can depend on the reaction temperature. This is because, although all reactions are reversible, it can be difficult to reach equilibrium and a non-equilibrium ratio of products can be obtained.

- At *low* temperatures, reactions are more likely to be *irreversible* and equilibrium is less likely to be reached. Under these conditions, the product that is formed at the fastest rate predominates. This is *kinetic control*. The *kinetic product* is therefore formed at the fastest rate (i.e. this product has the lowest activation energy barrier).
- At *high* temperatures, reactions are more likely to be *reversible* and an equilibrium is likely to be reached. Under these conditions, the energetically more stable product predominates. This is *thermodynamic control*. The *thermodynamic product* is therefore the most stable product (i.e. this product has the lowest energy).

The outcomes of reactions under kinetic control are determined by the relative energies of the transition states leading to various different products

The outcomes of reactions under thermodynamic control are determined by the relative energies of the various possible products

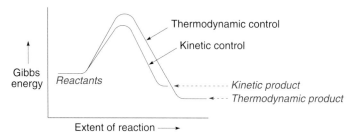

4.10 Orbital overlap and energy

Two atomic orbitals can combine to give two molecular orbitals – one *bonding* molecular orbital (lower in energy than the atomic orbitals) and one *antibonding* molecular orbital (higher in energy than the atomic orbitals) (Section 1.4). Orbitals that combine in-phase form a bonding molecular orbital and for best orbital overlap, the orbitals should be of the same size.

The orbitals can overlap end-on (as for σ-bonds) or side-on (as for π-bonds). The empty orbital of an electrophile (which accepts electrons) and the filled orbital of a nucleophile (which donates electrons) will point in certain directions in space. For the two to react, the filled and empty orbital must be correctly aligned; for end-on overlap, the filled orbital should point directly at the empty orbital.

The orbitals involved in an S_N2 reaction are shown in Section 5.3.1.1

Molecules must approach one another so that the filled orbital of the nucleophile can overlap with the empty orbital of the electrophile

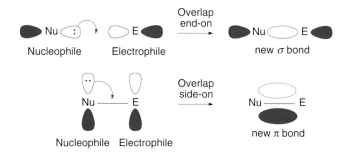

Acidity and basicity is discussed
in Section 1.7

Example: reaction of a base with HCl

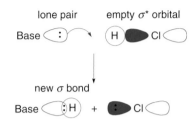

The orbitals must also have a similar energy. For the greatest interaction, the two orbitals should have the same energy. Only the *highest-energy occupied orbitals* (or HOMOs) of the nucleophile are likely to be similar in energy to only the *lowest-energy unoccupied orbitals* (or LUMOs) of the electrophile.

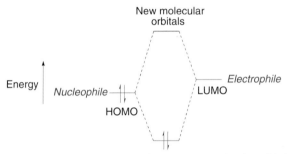

Strong nucleophiles have high energy HOMOs

The two electrons enter a lower energy molecular orbital.
There is therefore a gain in energy and a new bond is formed.
The further apart the HOMO and LUMO, the lower the gain in energy.

Strong electrophiles have low energy LUMOs.

- The HOMO of a nucleophile is usually a (non-bonding) lone pair or a (bonding) π-orbital. (These are higher in energy than a σ-orbital).
- The LUMO of an electrophile is usually an (antibonding) π^*-orbital. (This is lower in energy than a σ^* orbital.)

Example: addition of water to a ketone

For nucleophilic addition of water
to an aldehyde or ketone, to form a
hydrate, see Section 8.3.5.1

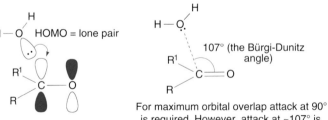

For maximum orbital overlap attack at 90°
is required. However, attack at ~107° is
observed because of the greater electron
density on the carbonyl oxygen atom,
which repels the lone pair on H_2O

4.11 Guidelines for drawing reaction mechanisms

1. Draw the structure of the reactants and include lone pairs on any heteroatoms. Show bond polarities by $\delta+$ and $\delta-$.
2. Identify which reactant is the nucleophile and which is the electrophile, and identify the nucleophilic atom in the nucleophile and the electrophilic atom in the electrophile.
3. Draw a double-headed curly arrow from the nucleophilic atom to the electrophilic atom to make a new bond. The arrow can start from a negative charge, a lone pair of electrons or a multiple bond. If a new bond is made to an uncharged H, C, N or O atom in the electrophile then one of the existing bonds should be broken and a second curly arrow should be drawn (pointing in the same direction as the first).
4. The overall charge of the reactants should be the same as that of the products. Try to end up with a negative charge on an electronegative atom.

> Take care and check that a carbon atom does not have more than four bonds

- If the *nucleophilic* atom is neutral then the atom will gain a positive charge, if it is negatively charged then it will become neutral.
- If the *electrophile* atom is neutral then one of the existing bonds must be broken, and a negative charge will reside on the most electronegative atom. If it is positively charged then it will become neutral.

> For clarity, it is often helpful to show curly arrows in a different colour to that used for the chemical structures

> For polar (ionic) reactions make sure that your curly arrows are double-headed; single headed arrows are used for radical reactions (Section 4.1)

> Both these reactions involve addition of a nucleophile to a ketone (see Section 8.3)

Worked example

The following questions relate to the reaction scheme shown below.

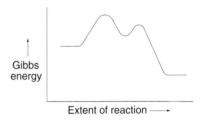

Hint: Consider the carbonyl carbon atoms

Hint: Compare the structures of **A** and **C**

Hint: Use double-headed curly arrows

(a) Use oxidation levels to help deduce if the conversion of **A** into **C** involves redox.

(b) Explain why the transformation of **A** into **C** is an example of a nucleophilic substitution reaction.

(c) Draw a reaction mechanism, using curly arrows, to show how **A** is converted into **B**, and how **B** is converted into **C**.

(d) The Gibbs energy profile for the reaction is shown below.

Hint: The energy minimum in the centre of the profile corresponds to a reaction intermediate

Hint: Consider the size of the activation energy for each step

Hint: The reaction is bimolecular

i. Draw the Gibbs energy profile and label it showing the location of the reactants, compound **B**, the products, and the transition states for Step 1 and Step 2. Also, add labels to show both $\Delta^{\ddagger}G$ and $\Delta_r G$.

ii. Giving your reasons, is Step 1 or Step 2 the rate-determining step?

iii. Experiments show that the rate equation for the reaction is:

$$\text{Rate of reaction} = k[\text{RO}^-][\mathbf{A}]$$

What is the overall order of the reaction?

Answer

Oxidation levels are discussed in Section 4.5

For general types of reaction see Section 4.6

(a) The oxidation levels of the carbonyl carbon atoms in **A** and **C** are both 3. So, the conversion of **A** into **C** involves neither oxidation nor reduction.

(b) The alkoxide ion (RO$^-$) is a nucleophile that reacts with compound **A** leading to substitution of the Cl atom in **A** for an RO group in **C**.

Drawing curly arrow mechanisms is discussed in Section 4.11

(c)

(d)   i.

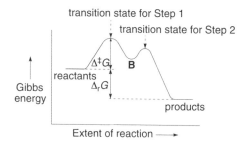

transition state for Step 1

transition state for Step 2

Gibbs energy profiles are discussed in Section 4.9.2

ii. Step 1 has the largest Gibbs energy of activation ($\Delta^{\ddagger}G$), so it is the slowest step, and the rate-determining step.

The rate-determining step is defined in Section 4.9.2.1

iii. The overall order of the reaction is two.

The order of a reaction is outlined in Section 4.9.2.1

Problems

1. Use oxidation levels to help deduce if the following reactions involve reduction or oxidation? If so, state whether the reaction involves oxidation or reduction of the organic reactants.

Section 4.5

2. For the following reactions, state which reactant can act as the nucleophile and which can act as the electrophile. Draw the likely product from each of the reactions and show its formation by using curly arrows.

Sections 4.2 and 4.11

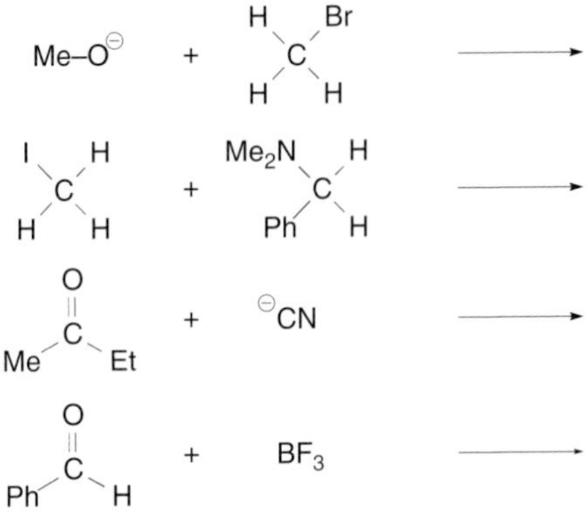

Section 4.6

3. Classify the following reactions as addition, elimination, substitution or rearrangement reactions.

(a) 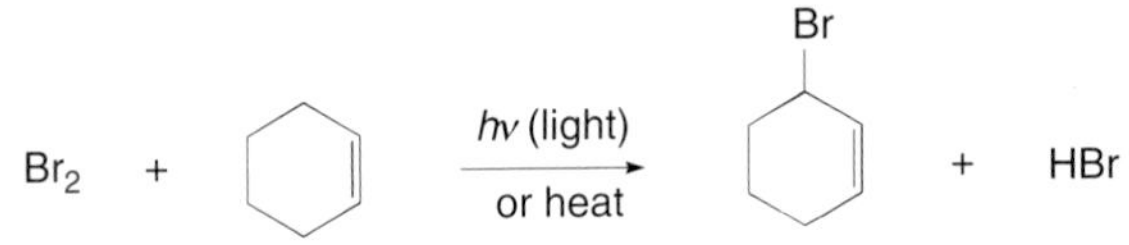   Br$_2$ + [cyclohexene] $\xrightarrow{hv\ (light)\ \text{or heat}}$ [bromocyclohexene] + HBr

(b) 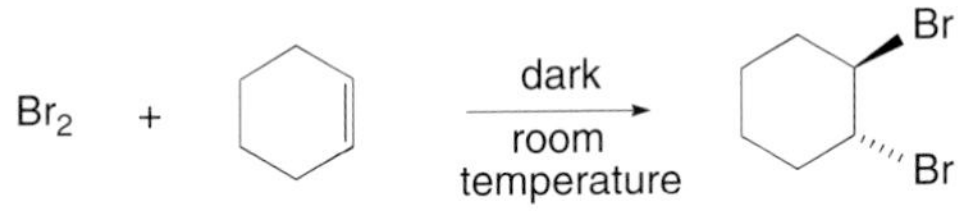   Br$_2$ + [cyclohexene] $\xrightarrow[\text{temperature}]{\text{dark room}}$ [dibromocyclohexane]

(c) 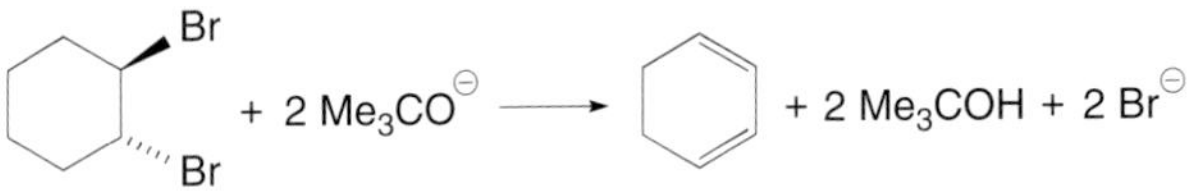   [dibromocyclohexane] + 2 Me$_3$CO$^{\ominus}$ $\longrightarrow$ [benzene] + 2 Me$_3$COH + 2 Br$^{\ominus}$

(d) 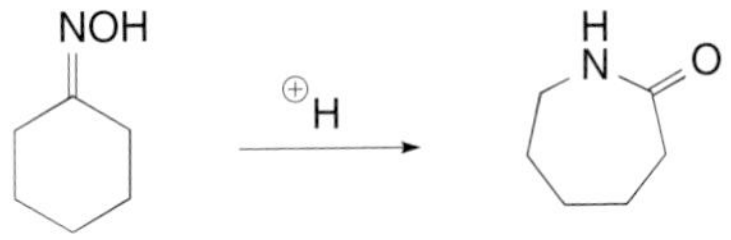   [cyclohexanone oxime] $\xrightarrow{\oplus H}$ [caprolactam]

Section 4.8

4. Classify the following reactions as either chemoselective, regioselective or stereoselective reactions.

(a) 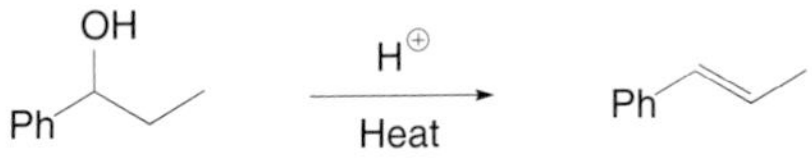   [Ph-CH(OH)-CH$_2$-CH$_3$] $\xrightarrow[\text{Heat}]{H^{\oplus}}$ [Ph-CH=CH-CH$_3$]

(b)    [2-bromobutane] $\xrightarrow{Me_3CO^{\ominus}}$ [1-butene]

(c) 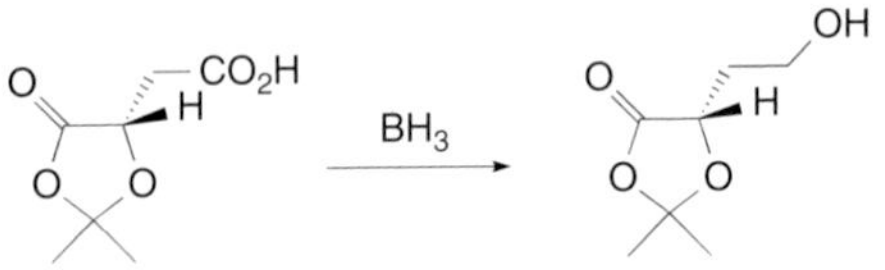   [dioxolane-CO$_2$H] $\xrightarrow{BH_3}$ [dioxolane-CH$_2$OH]

(d) 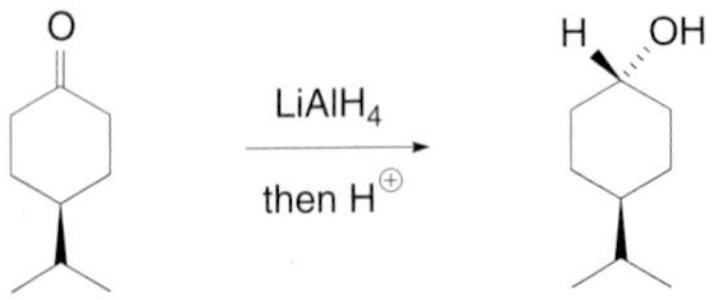   [4-isopropylcyclohexanone] $\xrightarrow[\text{then } H^{\oplus}]{LiAlH_4}$ [4-isopropylcyclohexanol]

5. The following questions relate to the reaction scheme shown below.

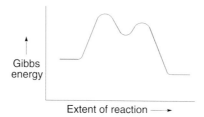

(a) Draw a reaction mechanism, using curly arrows, to show how **A** is converted into **B**, and how **B** is converted into **C**.   Section 4.11

(b) Would you expect carbocation **B** to be more or less stable than the ethyl cation ($CH_3CH_2^+$)? Explain your reasoning.   Sections 4.3 and 4.4

(c) Draw a diagram to show the orbital overlap in the reaction of carbocation **B** with hydroxide ion.   Section 4.10

(d) The Gibbs energy profile for the reaction is shown below.   Section 4.9

i. Draw the Gibbs energy profile and label it showing the location of the reactants, carbocation **B** and the products.

ii. Giving your reasons, is the reaction exothermic or endothermic?

iii. Giving your reasons, is the conversion of **A** to **B**, or the conversion of **B** to **C**, the rate-determining step?

6. The following questions are based on the reaction scheme shown below.

(a) Draw a reaction mechanism, using curly arrows, to show how **D** is converted into **G**.   Section 4.11

(b) Draw a diagram to show the orbitals that are involved when carbocation **D** reacts with HCl.   Section 4.10

(c) Provide an explanation for the rearrangement of carbocation **E** to form carbocation **F**.   Section 4.3.1

(d) Draw a diagram to show the orbitals that are involved when carbocation **E** reacts to form carbocation **F**.   Section 4.10

5

Halogenoalkanes

Key point. Halogenoalkanes (RX) are composed of an alkyl group bonded to a halogen atom (X = F, Cl, Br or I). As halogen atoms are more electronegative than carbon, the C–X bond is polar and nucleophiles can attack the slightly positive carbon atom. This leads to the halogen atom being replaced by the nucleophile in a *nucleophilic substitution reaction* and this can occur by either an S_N1 (*two-step*) or S_N2 (*concerted* or *one-step*) *mechanism*. In competition with substitution is *elimination*, which results in the loss of HX from halogenoalkanes to form alkenes. This can occur by either an *E1* (*two-step*) *or E2* (*concerted*) *mechanism*. The mechanism of the substitution or elimination reaction depends on the structure of the halogenoalkane, the solvent and the nucleophile/base.

5.1 Structure

Halogenoalkanes, R–X, have an alkyl group (R) joined to a halogen atom by a single bond. The larger the size of the halogen atom (X = I > Br > Cl > F), the weaker the C–X bond (Appendix 1). With the exception of C–I, the C–X bond is polar; the carbon atom bears a slight positive charge and the electronegative halogen atom bears a slight negative charge.

Keynotes in Organic Chemistry, Second Edition. Andrew F. Parsons.
© 2014 John Wiley & Sons, Ltd. Published 2014 by John Wiley & Sons, Ltd.

The use of hashed and wedged line notation is discussed in Section 3.3.2

$$\delta+ \quad \delta- \quad R = \text{alkyl or H}$$
$$R_3C \!-\! X \qquad X = F, Cl \text{ or } Br$$

Although the C–I bond is not polar, it becomes polarised when a nucleophile approaches it

sp3 hybridisation (tetrahedral)

$X = F, Cl, Br \text{ or } I$

Hybridisation is introduced in Section 1.5

Halogenoalkanes have an sp^3 carbon atom and so have a tetrahedral shape.

Naming halogenoalkanes is discussed in Section 2.4

The formation of bromobenzene from benzene is discussed in Section 7.2.1

- An *aliphatic* halogenoalkane has the carbon atoms in a chain and not a closed ring, e.g. 1-bromohexane, $CH_3(CH_2)_5Br$.
- An *alicyclic* halogenoalkane has the carbon atoms in a closed ring, but the ring is not aromatic, e.g. bromocyclohexane, $C_6H_{11}Br$.
- An *aromatic* halogenoalkane has the carbon atoms in a closed ring and the ring is aromatic, e.g. bromobenzene, C_6H_5Br.

5.2 Preparation

5.2.1 Halogenation of alkanes

Radicals are introduced in Section 4.1

Chloroalkanes (RCl) or bromoalkanes (RBr) can be obtained from alkanes (RH) by reaction with chlorine or bromine gas, respectively, in the presence of UV radiation. The reaction involves a radical chain mechanism.

UV radiation

Initiation $X \!-\! X \xrightarrow{\quad} 2X^\bullet$ $X = Cl \text{ or } Br$

Radical reactions are discussed in Section 4.6.2

reactant $\boxed{CH_4}$ $X^\bullet$ $\boxed{CH_3X}$ **product**

Propagation

HX X_2

$^\bullet CH_3$

Chloroalkanes, such as CH_2Cl_2 (dichloromethane), are common solvents in organic synthesis

This is a *substitution* reaction as a hydrogen atom on the carbon is substituted for a Cl or Br atom. A mixture of halogenated products is usually obtained if further substitution reactions can take place.

$$CH_4 \xrightarrow[]{X_2 \ HX} CH_3X \xrightarrow[]{X_2 \ HX} CH_2X_2 \xrightarrow[]{X_2 \ HX} CHX_3 \xrightarrow[]{X_2 \ HX} CX_4$$

Primary, secondary and tertiary halogenoalkanes are defined in Section 2.1

The ease of halogenation depends on whether the hydrogen atom is bonded to a primary, secondary or tertiary carbon atom. A tertiary hydrogen atom is more reactive because reaction with a halogen atom ($X^\bullet$) produces an intermediate tertiary radical, which is more stable (and therefore more readily formed) than a secondary or primary radical (Section 4.3).

Order of reactivity
toward X• R_3C—H > R_2CH—H > RCH_2—H

Order of radical $R_3\overset{\bullet}{C}$ > $R_2\overset{\bullet}{C}H$ > $R\overset{\bullet}{C}H_2$
stability   *tertiary*   *secondary*   *primary*

5.2.2 Halogenation of alcohols

Alcohols (ROH) are converted into halogenoalkanes using a number of methods. All methods involve 'activating' the OH group to make this into a better leaving group (Section 5.3.1.4).

Reaction mechanisms are introduced in Section 4.11

No substitution of OH for X

$X^\ominus$ + R—OH ⤫→ X—R + $^\ominus$OH

The hydroxide ion is relatively unstable and so is a poor leaving group

The mechanism of these reactions depends on whether a primary (RCH_2OH), secondary (R_2CHOH) or tertiary alcohol (R_3COH) is used. (Section 5.3.1 discusses the mechanisms.)

(i) Reaction with HX (X = Cl or Br)

pK_a values can be used as a guide for determining if a group is a good leaving group; a good leaving group has a low pK_a value (Section 5.3.1.4)

The OH group can be converted into a better leaving group, namely water, by protonation. As water is neutral, it is more stable than the hydroxide ion and so is a better leaving group.

- *Primary alcohol.* This proceeds via nucleophilic attack on an alkyloxonium ion (S_N2 mechanism).

protonation   concerted substitution reaction

$RCH_2\overset{..}{O}H$ + H—X ⟶ $X^\ominus$ RCH_2—$\overset{+}{O}$(H)(H) ⟶ X–CH_2R + H_2O
primary alcohol   *alkyloxonium ion*

In a concerted reaction, all bond making and bond breaking occurs in one step

- *Tertiary alcohol.* This proceeds via loss of water from the alkyloxonium ion (to give a relatively stable tertiary carbocation) before nucleophilic attack can take place (S_N1 mechanism).

Tertiary carbocations are stabilised by three electron-donating (+I) alkyl groups (Section 4.3.1)

protonation

R_3C–$\overset{..}{O}H$ + H—X ⟶ R_3C—$\overset{+}{O}$(H)(H) + $X^\ominus$
tertiary alcohol

stepwise substitution reaction   step 1 | –H_2O

R_3C–X ⟵ step 2 — $R_3\overset{+}{C}$ + $X^\ominus$
tertiary carbocation

- *Secondary alcohol.* This can proceed via either an S_N1 or S_N2 mechanism.

(ii) Reaction with phosphorus trihalides (PBr₃, PCl₃)

The OH group is converted into a neutral $HOPX_2$ leaving group (by an initial S_N2 reaction at phosphorus).

$P(OH)_3$, or H_3PO_3, is called phosphonic acid (or phosphorous acid)

Oxygen forms a strong bond with phosphorus

The $HOPX_2$ can react with two further moles of alcohol to form $P(OH)_3$

(iii) Reaction with thionyl chloride (SOCl₂)

Chloroalkanes can be prepared from reaction of alcohols (ROH) with thionyl chloride ($SOCl_2$) in the presence of a nitrogen base (e.g. triethylamine or pyridine). An intermediate alkyl chlorosulfite (ROSOCl) is formed by nucleophilic attack of ROH onto the sulfur atom of thionyl chloride. The OH group is converted into an OSOCl leaving group, which is displaced on reaction with the chloride anion (e.g. in an S_N2 mechanism when R is a primary alkyl group).

The reaction leads to the formation of a strong S=O bond at the expense of two weaker S–Cl bonds

A concerted (S_N2) reaction at the tetrahedral sulfur atom

alkyl chlorosulfite

The base 'mops up' the HCl by product

The evolution of SO_2 (gas) helps to drive the reaction to completion

Triethylamine (Et_3N) is a common base in organic synthesis (Section 1.7.2)

The mechanism changes to S_Ni in the absence of a nitrogen base (Section 5.3.1.7).

(iv) Reaction with p-toluenesulfonyl chloride

Chloroalkanes (RCl), bromoalkanes (RBr) and iodoalkanes (RI) can be formed from reaction of alcohols (ROH) with p-toluenesulfonyl chloride (or tosyl chloride, abbreviated TsCl) in the presence of a nitrogen base (e.g. triethylamine or pyridine). The OH group is converted into a tosylate (abbreviated ROTs), which can undergo substitution on reaction with Cl⁻, Br⁻ or I⁻. The stable tosylate

The influence of the alkyl group (R) group is discussed in Sections 5.3.1.2 and 5.3.1.2

anion ($^-$OTs) is an excellent leaving group (the substitution mechanism is S_N1 or S_N2 depending on the nature of the alkyl group, R).

tosyl chloride

Et$_3$N

concerted substitution reaction

+ Cl$^-$

The reaction leads to the formation of a strong S–O bond at the expense of a weaker S–Cl bond

Cl — R + $^-$O—S—CH$_3$ ← R—O—S—CH$_3$

tosylate anion   alkyl tosylate
                              +
                          Et$_3$N–H

The tosylate anion is an excellent leaving group because the negative charge can be stabilised by resonance. The charge is spread over the three electronegative oxygen atoms.

5.2.3 Halogenation of alkenes

Alkenes have an electron-rich C=C bond that can act as a nucleophile

The electrophilic addition of HX or X$_2$ to alkenes generates halogenoalkanes with one or two halogen atoms, respectively (see Section 6.2.2 for a detailed discussion of the reaction mechanisms).

Electrophilic addition reactions are introduced in Section 4.6.1

(i) Electrophilic addition of Br$_2$

The electron-rich alkene double bond repels the electrons in the bromine molecule to create a partial positive charge on the bromine atom near the double bond. An intermediate bromonium ion is formed, which reacts to give a dibromide derived from *anti* addition (i.e. the two Br groups add to the alkene from *opposite* sides). Consequently, reaction of a cycloalkene forms a *trans*-1,2-dibromocycloalkane.

A *trans*-1,2-dibromocycloalkane has the two Br atoms on opposite sides of the ring

E-alkene

cyclic bromonium ion

For assigning alkenes as *E* or *Z*, see Section 3.3.1.2

Opening of the 3-membered ring

Overall *anti* addition as the two bromine atoms land up on opposite sides of the planar alkene

The bromine atoms are on adjacent carbons

dibromoalkane

As the product with *anti*- stereochemistry is formed in excess over the *syn*-addition product (in which the two Br groups add to the alkene from the *same* side) the reaction is *stereoselective* (i.e. one particular stereoisomer of the product is formed in excess).

Reaction selectivity is introduced in Section 4.8

(ii) Electrophilic addition of HX

The electron-rich alkene double bond reacts with a proton (H^+), or with HX, so as to form the most stable intermediate carbocation. The addition is regioselective and the so-called Markovnikov (also spelled Markownikoff) product is formed. If a peroxide (ROOR) is added, the reaction gives the anti-Markovnikov product; the mechanism changes to one that involves radical intermediates (Section 6.2.2.1).

Carbocations are stabilised by electron-donating (+I) alkyl groups (Section 4.3.1)

trisubstituted alkene

The most stable carbocation is formed

H and X are on adjacent carbons

5.3 Reactions

Halogenoalkanes react with nucleophiles in *substitution* reactions and with bases in *elimination* reactions.

Substitution and elimination reactions are introduced in Section 4.6.1

5.3.1 Nucleophilic substitution

A good leaving group readily accepts a pair of electrons; as a guide, the weaker the basic strength of a group, the better leaving group it is

The two main mechanisms for nucleophilic substitution of halogenoalkanes (RX) are S_N1 and S_N2. These represent the extreme mechanisms of nucleophilic substitution and some reactions involve mechanisms which lie somewhere between the two.

In both S_N1 and S_N2 reactions, the mechanisms involve the loss of the halide anion (X^-) from RX. The halide anion that is expelled is called the *leaving group*.

5.3.1.1 The S_N2 (substitution, nucleophilic, bimolecular) reaction

This is a concerted (one-step) mechanism in which the nucleophile forms a new bond to carbon at the same time as the bond to the halogen atom (X) is broken. The reaction is second order, or bimolecular, as the rate depends on the concentration of both the nucleophile (Nu) and the halogenoalkane (RX).

Reaction rates are discussed in Section 4.9.2.1

$$\text{Reaction rate} = k[\text{Nu}][\text{RX}]$$

The reaction leads to an inversion (or change) of stereochemistry at a chiral centre (i.e. an *R*-enantiomer will be converted into an *S*-enantiomer). This is known as a *Walden inversion*.

For assigning *R* and *S* configuration, see Section 3.3.2.2

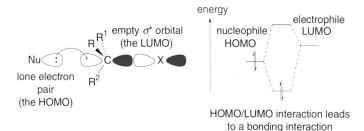

Nucleophile approaches at an angle of 180° to the C–X bond which is broken   Transition State (T.S.)   Inversion of configuration

The charge is spread from the nucleophile (Nu) to the leaving group (X). The carbon atom in the T.S. is partially bonded to both Nu and X

The nucleophile approaches the C–X bond at an angle of 180°. This maximises the interaction of the filled orbital of the nucleophile with the empty $\sigma*$ orbital of the C–X bond.

Orbital overlap is discussed in Section 4.10

HOMO stands for highest energy occupied orbital

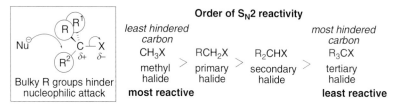

HOMO/LUMO interaction leads to a bonding interaction

LUMO stands for lowest energy unoccupied orbital

Halogenoalkanes with bulky alkyl substituents react more slowly than those with small alkyl substituents on the central carbon atom. Bulky substituents prevent the nucleophile from approaching the central carbon atom. S_N2 reactions can therefore only occur at relatively unhindered sites.

Steric effects and steric hindrance is discussed in Section 4.4

Order of S_N2 reactivity

	least hindered carbon			most hindered carbon
	CH_3X	RCH_2X	R_2CHX	R_3CX
	methyl halide	primary halide	secondary halide	tertiary halide
	most reactive			**least reactive**

Bulky R groups hinder nucleophilic attack

> methyl halide > primary halide > secondary halide > tertiary halide

The terms primary, secondary and tertiary are introduced in Section 2.1

Inductive effects are introduced in Section 1.6.1

The presence of +I alkyl groups on the central carbon atom also reduces the partial positive charge, which reduces the rate at which the nucleophile attacks the carbon atom.

5.3.1.2   *The S_N1 (substitution, nucleophilic, unimolecular) reaction*

This is a stepwise mechanism involving initial cleavage of the carbon–halogen bond to form an intermediate carbocation. The reaction is first order, or unimolecular, as the rate depends on the concentration of only the halogenoalkane (and *not* the nucleophile).

Reaction rates are discussed in Section 4.9.2.1

$$\text{Reaction rate} = k[\text{RX}]$$

The reaction leads to racemisation of a chiral centre in the starting material (i.e. an *R*-enantiomer will be converted into a 50:50 mixture of *R*- and *S*-enantiomers). This is because the nucleophile can equally attack either side of the planar carbocation.

For assigning *R* and *S* configuration, see Section 3.3.2.2

Carbocations are planar and contain an empty p orbital (Section 4.3)

Initial C–X bond cleavage in a slow, rate-determining step

Intermediate planar carbocation

Equal amounts of enantiomers (a racemate)

The more stable the carbocation intermediate, the faster the S_N1 reaction (i.e. the easier it is to break the C–X bond). Tertiary halides (R_3CX) will therefore react faster than primary halides (RCH_2X) by this mechanism because a tertiary carbocation (R_3C^+) is more stable than a primary carbocation (RCH_2^+).

The stability of carbocations is discussed in Section 4.3.1

Order of S_N1 reactivity

produces most stable carbocation

R_3CX
tertiary halide

R_2CHX
secondary halide

RCH_2X
primary halide

CH_3X
methyl halide

produces least stable carbocation

most reactive   **least reactive**

Stabilisation of cations by resonance (involving +M groups) is discussed in Section 1.6.3

Primary benzylic and allylic halides can undergo S_N1 reactions because the carbocations are stabilised by resonance. These carbocations have similar stability to secondary alkyl carbocations.

Bn is commonly used as an abbreviation for a benzyl group, e.g. BnBr represents benzyl bromide ($C_6H_5CH_2Br$)

5.3.1.3 *S_N2 versus S_N1 reactions*

The mechanism of a nucleophilic substitution reaction is influenced by the nature of the halogenoalkane, the nucleophile and the solvent.

The halogenoalkane

- *Primary* halogenoalkanes are likely to react by an S_N2 mechanism.
- *Secondary* halogenoalkanes are likely to react by either an S_N1 or S_N2 mechanism (or an intermediate pathway with both S_N1 and S_N2 character).
- *Tertiary* halogenoalkanes are likely to react by an S_N1 mechanism.

The nucleophile

- Increasing the nucleophilic strength of the nucleophile *will* increase the rate of an S_N2 reaction.
- Increasing the nucleophilic strength of the nucleophile *will not* increase the rate of an S_N1 reaction.

In an S_N2 reaction, the nucleophile takes part in the rate-determining step; strong nucleophiles include the iodide ion (Section 5.3.1.5)

The mechanism can shift from S_N1 to S_N2 on changing to a stronger nucleophile.

The solvent

- Increasing the polarity (i.e. the relative permittivity, ε_r) of the solvent will result in a *slight decrease* in the rate of an S_N2 reaction. This is because the charge is more spread out in the transition-state than in the reactants.
- Increasing the polarity of the solvent will result in a *significant increase* in rate of an S_N1 reaction. This is because polar solvents, with high relative permittivity values (e.g. water and methanol), can stabilise the carbocation intermediate (formed in an S_N1 reaction) by solvation. For an S_N2 reaction, solvation is likely to stabilise the (more concentrated negative charge of the) attacking nucleophile rather than the (less concentrated charge which is spread over the) transition state. The mechanism can therefore shift from S_N2 to S_N1 on changing to a more polar solvent.
- In addition, the nucleophilic strength of a nucleophile depends on whether a polar protic solvent or a polar non-protic (or aprotic) solvent is used. This is because polar protic solvents (e.g. methanol) can stabilise, and therefore lower the reactivity of the attacking nucleophile, by hydrogen bonding. The nucleophilic strength is increased in polar non-protic solvents (e.g. dimethyl sulfoxide, Me_2SO), because these are not as effective at solvating the nucleophile (i.e. they are not able to hydrogen bond to a negatively charged nucleophile). The mechanism can therefore shift from S_N1 to S_N2 on changing from a polar protic solvent to a polar non-protic solvent.

An older name for relative permittivity is dielectric constant

The ε_r values for water, methanol and hexane are 79, 33 and 2, respectively

Polar protic solvents contain O–H or N–H bonds that form hydrogen bonds to anions and cations

Hydrogen bonding is introduced in Section 1.1

5.3.1.4   The leaving group

The better the leaving group, the faster the rate of *both* S_N1 and S_N2 reactions (as C–X bond cleavage is involved in both of the rate determining steps). The best leaving groups are those that form stable neutral molecules or stable anions,

H_3O^+ is a stronger acid than H_2O, so H_2O is a better leaving group than HO^-

which are weakly basic. The less basic the anion, the more stable (or less reactive) the anion.

Leaving group ability

$$H_2O \quad > \quad HO^{\ominus}$$

$$I^{\ominus} \quad > \quad Br^{\ominus} \quad > \quad Cl^{\ominus} \quad > \quad F^{\ominus}$$

weak base   strong base

HI is a stronger acid than HF

Amongst the halogens, the iodide anion (I^-) is the best leaving group as it forms a weak bond to carbon (Appendix 1). The C–I bond is therefore relatively easily broken to give I^-. I^- is a weak base and weak bases are best able to accommodate the negative charge (i.e. weak bases are the best leaving groups).

5.3.1.5 Nucleophilic catalysis

A catalyst increases the rate of a reaction by allowing the reaction to proceed by a different pathway, which has a lower energy transition state (Section 4.9.2)

The iodide anion (I^-) is a good nucleophile *and* a good leaving group. It can be used as a catalyst to speed up the rate of a slow S_N2 reaction.

5.3.1.6 Tight ion pairs

In practice, few S_N1 reactions lead to complete racemisation because of the formation of *tight* (or *intimate*) *ion pairs*. The carbocation and negatively charged leaving group interact so that the negatively charged nucleophile enters predominantly from the opposite side to the departing leaving group. This results in a greater proportion of inversion of configuration.

tight ion pair

5.3.1.7 The $S_N i$ reaction (substitution, nucleophilic, internal reaction)

When an ion (or molecule) is surrounded by solvent molecules (that interact by non-covalent interactions), the solvent molecules are called the solvent shell

The formation of a tight (or intimate) ion pair can also lead to retention of configuration. This is observed when alcohols (ROH) with chiral centres react with thionyl chloride ($SOCl_2$) in the absence of a nitrogen base. The intermediate alkyl chlorosulfite (ROSOCl) breaks down to form a tight ion pair ($R^{+-}OSOCl$) within a solvent shell. The ^-OSOCl anion then collapses to form SO_2 and Cl^-; the Cl^- then attacks the carbocation from the same side that ^-OSOCl departs.

tertiary alcohol

alkyl chlorosulfite

Attack by Cl$^{\ominus}$ occurs on the same side as $^{\ominus}$OSOCl departs

tertiary carbocation

tight ion pair

Solvent shell surrounds the ions

Primary, secondary and tertiary alcohols are introduced in Section 2.1

- In the *absence* of a nitrogen base, HCl is formed and this cannot act as a nucleophile and attack the alkyl chlorosulfite in an S_N2 reaction. Therefore, an S_Ni mechanism operates.
- An S_N2 reaction is observed in the *presence* of a nitrogen base (B; such as triethylamine)) because the HCl is converted into BH$^+$ Cl$^-$. The Cl$^-$ can then act as a nucleophile and attack the alkyl chlorosulfite in an S_N2 reaction (see Section 5.2.2).

Et$_3$N and related bases are discussed in Section 1.7.2

5.3.1.8 Neighbouring group participation (or anchimeric assistance)

Two consecutive S_N2 reactions will lead to a retention of configuration (i.e two inversions of configuration = retention). This can occur when a neighbouring group acts as a nucleophile in the first of two S_N2 reactions.

- The *first* S_N2 reaction is an *intramolecular* reaction (i.e. a reaction within the same molecule).
- The *second* S_N2 reaction is an *intermolecular* reaction (i.e. a reaction between two different molecules).

X is a leaving group

Base$^{\ominus}$
(− BaseH)

First S_N2
(intramolecular)
(− X$^{\ominus}$)

Second S_N2
(intermolecular)
then H$^{\oplus}$

The carbon atoms adjacent to O are electrophilic

epoxide

For the formation of epoxides from alkenes, see Section 6.2.2.6

The 3-membered epoxide ring is highly strained and reacts with various nucleophiles

The presence of neighbouring groups, including R_2N and RS, can increase the rate of substitution reactions.

Compounds of type R_2S are called sulfides, or thioethers (Section 2.1)

First S_N2 $(-X^\ominus)$ Second S_N2

The carbon atoms adjacent to S are electrophilic
(the positively charged sulfur has a –I effect)

–I effects are introduced in Section 1.6.1

5.3.1.9   The S_N2' and S_N1' reactions

Allylic halides can undergo substitution with an allylic rearrangement (i.e. change in position of the C=C bond) in one of two ways.

The allyl group is $H_2C=CH-CH_2-$

S_N2' mechanism: by nucleophilic attack on the double bond

concerted mechanism

S_N2'

allylic halide(X = halogen leaving group)

S_N1' mechanism: by nucleophilic attack at a carbocation resonance structure

stepwise mechanism

$-X^\ominus$

primary carbocation

Stabilisation of carbocations by resonance is discussed in Section 1.6.3

carbocation stabilised by resonance

secondary carbocation

5.3.2   Elimination

Elimination reactions are introduced in Section 4.6.1.2

The two main mechanisms for elimination reactions of halogenoalkanes are E1 and E2. In both cases, the mechanism involves the loss of HX from RX (e.g. $R_2CH-CXR_2$) to form an alkene (e.g. $R_2C=CR_2$).

5.3.2.1   The E2 (elimination, bimolecular) reaction

This is a concerted (one-step) mechanism. The C=C bond begins to form at the same time as the C–H and C–X bonds begin to break. The reaction is second order (or bimolecular) as the rate depends on the concentration of the base and the halogenoalkane (RX).

$$\text{Reaction rate} = k[\text{Base}][\text{RX}]$$

The elimination requires the halogenoalkane to adopt an *antiperiplanar* shape (or conformation), in which the H and X groups are on opposite sides of the molecule. (*Synperiplanar* is when the H and X groups are on the same side of the molecule.) The antiperiplanar arrangement is lower in energy than the synperiplanar arrangement as this has a staggered, rather than an eclipsed, conformation.

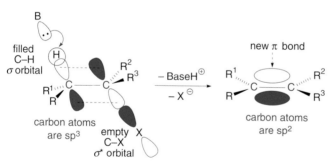

The elimination is *stereospecific* as different stereoisomers of the halogenoalkane give different stereoisomers of the alkene. This is because the new π-bond is formed by overlap of the C−H σ-orbital with the C−X σ^*-orbital. These orbitals must be in the same plane for the best overlap and they become p orbitals in the π-bond of the alkene.

The antiperiplanar conformation is responsible for:

i. the stereospecific formation of substituted alkenes (i.e. alkenes of *E* or *Z* configuration);
ii. the different rates of elimination of HX from halocyclohexane conformers.

(i) Elimination reactions of halogenoalkane diastereoisomers

A different alkene stereoisomer is obtained from each diastereoisomer of the halogenoalkane.

Staggered and eclipsed conformations are discussed in Section 3.2.1

The double dagger ($\ddagger$) indicates that this is a transition state

sp^3 and sp^2 hybridisation is discussed in Section 1.5

Assignment of *E* and *Z* configuration is discussed in Section 3.3.1.2

Diastereoisomers are stereoisomers that are not mirror images of each other (see Section 3.3.2.4)

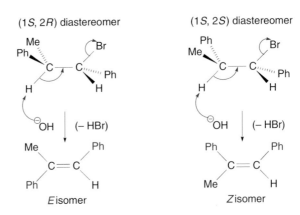

(1S, 2R) diastereomer   (1S, 2S) diastereomer

For an *E*-alkene, the groups of highest priority are on the opposite sides of the C=C bond (Section 3.3.1.2)

E isomer   *Z* isomer

(ii) Elimination reactions of halocyclohexane conformers

The chair conformation of cyclohexane, together with axial and equatorial bonds, is discussed in Section 3.2.4

For elimination in halocyclohexanes both the C−H and C−X bonds must be axial.

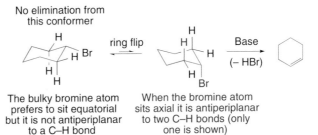

No elimination from this conformer

ring flip   Base
(− HBr)

The bulky bromine atom prefers to sit equatorial but it is not antiperiplanar to a C−H bond

When the bromine atom sits axial it is antiperiplanar to two C−H bonds (only one is shown)

Regioselective reactions are introduced in Section 4.8

(iii) Regioselectivity

The regioselectivity (i.e. the position of the C=C bond in the alkene) of the E2 reaction depends on the nature of the leaving group (X).

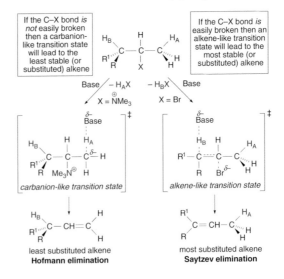

If the C−X bond *is not easily broken* then a carbanion-like transition state will lead to the least stable (or substituted) alkene

If the C−X bond *is easily broken* then an alkene-like transition state will lead to the most stable (or substituted) alkene

Base $\diagup$ − H$_A$X   − H$_B$X $\diagdown$ Base
X = $\overset{\oplus}{N}$Me$_3$ X = Br

Compounds of type R−NR$_3^+$ are called quaternary ammonium ions (Section 2.1)

carbanion-like transition state   alkene-like transition state

least substituted alkene
Hofmann elimination

most substituted alkene
Saytzev elimination

- *When X = Br.* The elimination reaction forms predominantly the *most* highly substituted alkene in a *Saytzev* (or *Zaitsev*) *elimination*. The H–C bond is broken at the same time as the C–Br bond and the transition state resembles the alkene product. As a result, the more highly substituted alkene is formed faster because this is the more stable alkene product (Section 6.1). An alkene is stabilised by alkyl substituents on the C=C bond

- *When X = $^+NMe_3$.* The elimination reaction forms predominantly the *least* highly substituted alkene in a *Hofmann elimination*. The H–C bond is broken before the C–$^+$NMe$_3$ bond and the transition state resembles a carbanion product. As a result, the base removes H^A to form a more stable carbanion (i.e. the $\delta-$ is situated on a carbon bearing the least number of electron-donating alkyl groups). The Hofmann elimination is also normally expected when using bulky bases (e.g. ^{t}BuO$^-$). This is because H^A is more accessible (i.e. less hindered and more easily attacked) than H^B. The stability of carbanions is discussed in Section 4.3.1

5.3.2.2 The E1 (elimination, unimolecular) reaction

Orders of reaction and rate equations are discussed in Section 4.9.2.1

This is a stepwise mechanism involving initial cleavage of the carbon–halogen bond (R–X) to form an intermediate carbocation (R$^+$). The reaction is first order, or unimolecular, as the rate depends on the concentration of only the halogenoalkane (and not the base).

$$\text{Reaction rate} = k[\text{RX}]$$

The reaction does not require a particular geometry; the intermediate carbocation can lose any proton from an adjacent carbon.

Sometimes the base is not included and H$^+$ is shown dropping off – however, a base is always needed to remove the H$^+$

Tertiary halogenoalkanes (R$_3$CX) react more rapidly than primary halogenoalkanes (RCH$_2$X). This is because the intermediate tertiary carbocation (R$_3$C$^+$) is more stable than a primary carbocation (RCH$_2^+$), and therefore it is more readily formed. The stability of carbocations is discussed in Section 4.3.1

Order of E1 reactivity:

$$\text{tertiary}(R_3CX) > \text{secondary}(R_2CHX) > \text{primary}(RCH_2X)$$

E1 reactions can be regioselective and stereoselective. Regioselective and stereoselective reactions are introduced in Section 4.8

(i) Regioselectivity

E1 eliminations give predominantly the more stable (or more substituted) alkene. The transition state (for loss of a proton from the intermediate carbocation) leading to the more substituted alkene will therefore be lower in energy.

(ii) Stereoselectivity

Assignment of *E* and *Z* configuration is discussed in Section 3.3.1.2

E1 eliminations give predominantly the *E*-alkene rather than the *Z*-alkene. The *E*-alkene is more stable than the *Z*-alkene for steric reasons (i.e. the bulky substituents are further apart in *E*-alkenes). The transition state leading to *E*-alkenes will therefore be lower in energy.

Reaction intermediates are often shown in square brackets

major product = most substituted alkene with the *E*-configuration

C–C bond rotation

In this example, the C=C bond is stabilised by conjugation with the adjacent benzene ring (Section 1.6.3)

removal of a proton from this carbon would give a disubstituted alkene

trisubstituted *E*-alkene

Bulky Et and Ph groups are on opposite sides

5.3.2.3 E2 versus E1 reactions

The mechanism of an elimination reaction is influenced by the nature of the halogenoalkane, the base and the solvent.

The halogenoalkane

The rate-determining step is the slowest step of a chemical reaction (Section 4.9.2.1)

- Primary (RCH_2X), secondary (R_2CHX) and tertiary (R_3CX) halogenoalkanes can all undergo an E2 mechanism (hence this is more common than E1).
- Tertiary halogenoalkanes (R_3CX) can react by an E1 mechanism.
- The better the leaving group (X), the faster the rate of both E1 and E2 reactions (as C–X bond cleavage is involved in both of the rate determining steps).

The base

Acid-base reactions are discussed in Section 1.7.6

- Increasing the strength of the base *will* increase the rate of an E2 elimination. Strong bases such as ${}^tBuO^-$ and ${}^iPr_2N^-$ will favour an E2 elimination.
- Increasing the strength of the base *will not* increase the rate of an E1 reaction.
- The mechanism can shift from E1 to E2 on changing to a more powerful base.

${}^iPr_2NLi$ or $(Me_2CH)_2NLi$, is called lithium diisopropylamide; it is a strong base, commonly used in synthesis

The solvent

- Increasing the polarity (i.e. the relative permittivity, ε_r) of the solvent will result in a *slight decrease* in the rate of an E2 reaction.
- Increasing the polarity of the solvent will result in a *significant increase* in the rate of an E1 reaction.

This is because polar solvents, with high relative permittivities (e.g. water and methanol), can stabilise the carbocation intermediate (formed in an E1 reaction) by solvation. For an E2 reaction, solvation is likely to stabilise the attacking base (which has a more concentrated charge) rather than the transition state (which has a less concentrated charge). The mechanism can therefore shift from E2 to E1 on changing to a more polar solvent.

- The strength of a base depends on whether a polar protic solvent or a polar non-protic (or aprotic) solvent is used. This is because polar protic solvents (e.g. methanol) can stabilise, and therefore lower the reactivity of the attacking base, by hydrogen bonding. The basic strength is increased in polar non-protic solvents (e.g. dimethyl sulfoxide, Me_2SO), because these are not as effective at solvating the base (i.e. they are not able to hydrogen bond to a negatively charged base). The mechanism can therefore shift from E1 to E2 on changing from a polar protic solvent to a polar non-protic solvent.

Attraction of molecules of a solvent to molecules or ions of a solute is called solvation

Hydrogen bonding is the interaction of a hydrogen atom in one molecule with an electronegative atom, such as oxygen, in another molecule (Section 1.1)

5.3.2.4 The ElcB (elimination, unimolecular, conjugate base) reaction

This reaction involves a stepwise mechanism that starts by deprotonation of the halogenoalkane to form an intermediate carbanion. In a second step, the carbanion (or conjugate base of the starting material) then loses the leaving group (X^-) in a slow (rate-determining) step.

The rate-determining step is the slowest step of a chemical reaction (Section 4.9.2.1)

Acid-base reactions are discussed in Section 1.7.6

The reaction is favoured by the presence of electron withdrawing ($-I$, $-M$) groups on the β-carbon atom (i.e. R and R^1 = electron withdrawing groups). Suitable groups include $RC=O$ and RSO_2, which can stabilise the intermediate carbanion. These groups therefore increase the acidity of the β-hydrogen atoms leading to an initial deprotonation.

$-I$ and $-M$ groups are introduced in Sections 1.6.1 and 1.6.3

5.3.3 Substitution versus elimination

Halogenoalkanes undergo competitive substitution and elimination reactions. The ratio of products derived from substitution and elimination depends on the structure of the halogenoalkane, the choice of base or nucleophile, the reaction solvent and the temperature. S_N2 reactions are normally in competition with E2 reactions, while S_N1 reactions are normally in competition with E1 reactions.

(i) S_N2 versus E2

Primary (RCH_2X) and secondary (R_2CHX) halogenoalkanes can undergo S_N2 reactions while primary (RCH_2X), secondary (R_2CHX) and tertiary (R_3CX) halogenoalkanes can undergo E2 reactions.

For primary and secondary halogenoalkanes

- S_N2 reactions are favoured over E2 reactions when using strong nucleophiles in polar aprotic solvents.
- E2 reactions are favoured over S_N2 reactions by using strong bulky bases (which are poor nucleophiles).

Factors that influence the strength of nucleophiles and bases are discussed in Sections 4.2 and 1.7.2

In general, large nucleophiles are good bases and promote elimination. This is because the bulky anion cannot attack the hindered carbon atom in an S_N2 reaction. It is much easier for the anion to attack a β-hydrogen atom because this is more accessible.

When the size of groups is responsible for reducing the reactivity at a site within a molecule, this is called steric hindrance (Section 4.4)

With a large anion, this is more likely to act as a base and react with the hydrogen on the β-carbon because this is more exposed than the α-carbon atom

Δ_rG^θ is the standard Gibbs free-energy change of reaction

E2 reactions are favoured over S_N2 reactions by using high temperatures. In general, increasing the reaction temperature leads to more elimination. This is because the elimination reaction has a higher activation energy (than the substitution reaction) as more bonds need to be broken in order to form the alkene product.

It should also be noted that in an elimination reaction, 2 molecules react to give 3 new product molecules. In contrast, for a substitution reaction, 2 molecules react to form 2 new product molecules. The change in entropy (Δ_rS^θ) is therefore greater for an elimination reaction. From the Gibbs free-energy equation (Section 4.9.1), Δ_rS^θ is multiplied by the temperature (T) and the larger the $T\Delta_rS^\theta$ term, the more favourable Δ_rG^θ.

(ii) S_N1 versus E1

Allylic and benzylic carbocations are stabilised by resonance (Section 1.6.3)

Secondary (R_2CHX) and tertiary (R_3CX) halogenoalkanes can undergo S_N1 or E1 reactions. Allylic (e.g. $H_2C=CHCH_2X$) and benzylic (e.g. $PhCH_2X$) halides can also undergo S_N1 or E1 reactions. (Tertiary (R_3CX) halogenoalkanes usually undergo E2 elimination in the presence of strong bases.) Both S_N1 and E1 reactions can occur when using weakly basic or non-basic nucleophiles in protic solvents.

E1 reactions are favoured over S_N1 reactions by using high temperatures. E1 reactions are also favoured as the size of the alkyl groups (on the α-carbon atom) increases. The bigger these groups are, the harder it is for the nucleophile to attack the α-carbon atom.

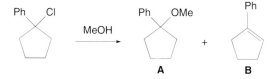

The bigger the alkyl groups the more likely the anion will attack the hydrogen atom on the β-carbon

(iii) S_N2/E2 versus S_N1/E1

- S_N1/E1 reactions are favoured over S_N2/E2 reactions by using polar protic solvents (which can solvate the carbocation intermediates).
- S_N2/E2 reactions are favoured over S_N1/E1 reactions by using strong nucleophiles or bases.
- S_N2/E2 reactions are favoured over S_N1/E1 reactions by using high concentrations of the nucleophile/base (as the rate of these bimolecular reactions depend on the concentration of the nucleophile or base).

Methanol (CH_3OH) is a commonly used polar protic solvent; dimethylformamide (DMF, Me_2NCHO) is a common polar aprotic solvent

Halogenoalkane	Mechanism
Primary: RCH_2X	S_N2, *E2*
Secondary: R_2CHX	S_N1, S_N2, E1, *E2*
Tertiary: R_3CX	S_N1, E1, *E2*

In *italic* = favoured when using a strong base

In blue = favoured when using a good nucleophile

Sometimes, 1°, 2° and 3° is used to represent primary, secondary and tertiary (Section 2.3)

Worked example

(a) Giving your reasons, explain which reaction mechanism (S_N1, S_N2, E1 or E2) reactions (i) and (ii) follow, and draw the structure of the major product.

Hint: Identify the leaving group and the base or nucleophile. Consider the strength of the base or nucleophile and the type of solvent

(i) [structure: chlorocyclopentane] Cl
$\xrightarrow{\ominus CN}{Me_2SO}$

(ii) [structure: bromo-dimethyl cyclopentane] Br
$\xrightarrow{(CH_3)_3C-O^\ominus}{Me_2NCHO}$

(b) Draw reaction mechanisms to explain the formation of products **A** and **B** below.

Hint: Decide if the products are formed by a substitution or an elimination reaction and consider whether methanol is a polar protic or aprotic solvent

Ph Cl [cyclopentane structure] $\xrightarrow{MeOH}$ Ph OMe [cyclopentane structure] + Ph [cyclopentene structure]

A   **B**

Answer

(a) i. This reaction is an S_N2 reaction. The cyanide ion ($^-$CN) is a strong nucleophile and Me$_2$SO is a polar aprotic solvent.

The S_N2 reaction is discussed in Section 5.3.1.2

ii. This reaction is an E2 reaction. The *tert*-butoxide ion (Me$_3$CO$^-$) is a strong base and Me$_2$NCHO is a polar aprotic solvent. A disubstituted alkene is formed because the bulky Me$_3$CO$^-$ ion removes a proton from the least hindered β-carbon.

The E2 reaction is discussed in Section 5.3.2.1

(b) Both the formation of **A** and **B** involve the formation of an intermediate benzylic cation, which is stabilised by resonance.

S_N1 and E1 reactions are introduced in Sections 5.3.1.2 and 5.3.2.2

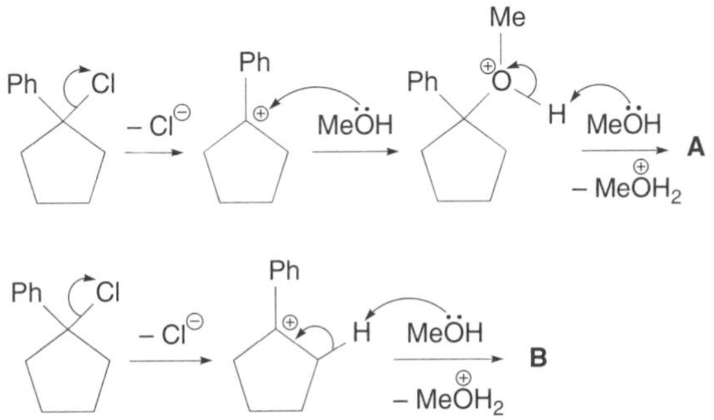

Problems

1. For the reaction of compound **A** with bromide ion in propanone (acetone):

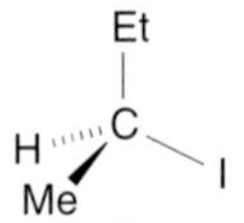

Section 5.3.1.1

Section 5.3.1.3

Section 5.3.3

(a) Draw the product of an S_N2 reaction.
(b) Predict what other substitution product might be formed if the reaction is carried out in ethanol instead of propanone.
(c) Explain why, if **A** is reacted with hydroxide ion rather than bromide ion, little substitution product results.

(d) Explain why, in an S_N1 reaction, compound **A** reacts less rapidly than does (1-iodoethyl)benzene, $PhCH(I)CH_3$.   Section 5.3.1.2

2. Hydrolysis of 2-bromopentane (**B**) can occur by either an S_N1 pathway or by an S_N2 pathway, depending on the solvent and conditions employed.   Section 5.3.1.2

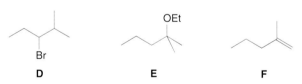

(a) Draw the structure of the (*R*)-isomer of **B**.   Section 3.3.2.2

(b) Write down the S_N1 and S_N2 mechanism of the above reaction, starting from the (*R*)-isomer of **B**.   Sections 5.3.1.1 and 5.3.1.2

(c) Suggest, giving your reasons, conditions under which the reaction might be expected to proceed by an S_N2 route.   Section 5.3.1.3

(d) Explain why addition of catalytic amounts of sodium iodide accelerates the formation of **C** from **B** under S_N2 but not S_N1 conditions.   Section 5.3.1.5

(e) What product(s) would you expect from reaction of **B** with sodium *tert*-butoxide, $Me_3C–O^-$ Na^+?   Section 5.3.2.1

(f) Explain why **C** is stable in ethanol, but if a small amount of sulfuric acid is added, it is slowly converted into 2-ethoxypentane.   Section 5.2.2

3. When the solvent acts as the nucleophile in a substitution reaction, this is known as *solvolysis*. What products would you expect from solvolysis of $PhCH_2Br$ in (a) methanol and (b) ethanoic acid?   Sections 5.3.1.1 and 5.3.1.2

4. When (*R*)-butan-2-ol is reacted with excess $SOCl_2$, the main product is (*R*)-2-chlorobutane. However, if the same is carried out in the presence of pyridine, the major product is (*S*)-2-chlorobutane. Why do the two reactions give different products?   Section 5.3.1.7

5. Draw the structure of the product formed on E2 elimination of (1*S*,2*S*)-1,2-dichloro-1,2-diphenylethane.   Section 5.3.2.1

6. Reaction of bromide **D** in hot ethanol gave a number of products including **E** and **F**. Suggest reaction mechanisms to explain the formation of **E** and **F**.   Sections 5.3.3 and 4.6.1.4

OEt

D   **E**   **F**

7. (a) Propose reaction mechanisms to explain the formation of compounds **G** and **H**, which involves an intermediate enolate ion.

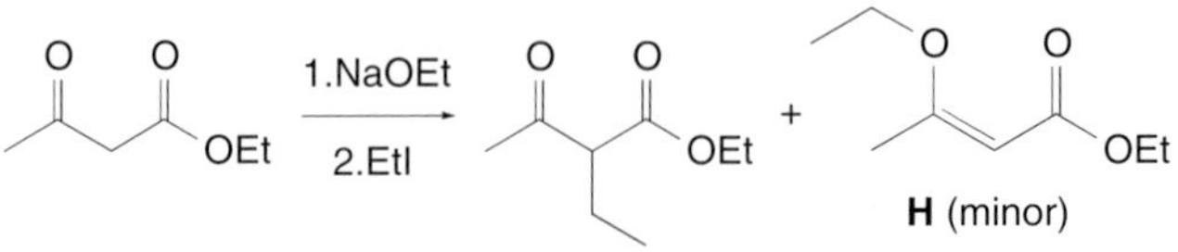

Sections 1.7.6 and 5.3.1.1

G (major)

H (minor)

Sections 5.3.1.9 and 5.3.2.4

(b) Draw a reaction mechanism, that does not involve carbocation intermediates, to explain how compound **I** is formed.

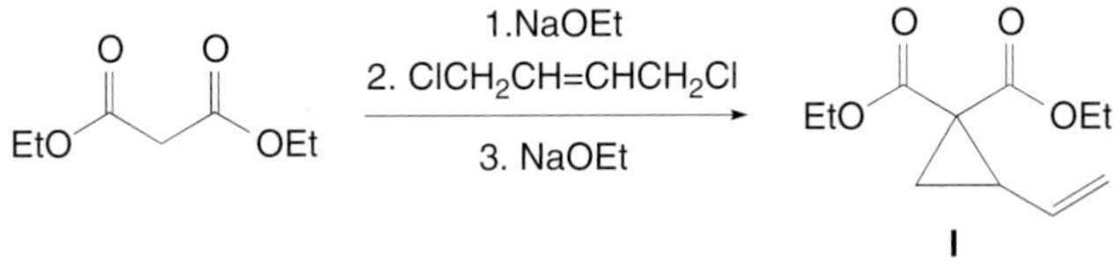

I

Sections 1.7.6 and 5.3.1.1

(c) Draw the structure of product **J**.

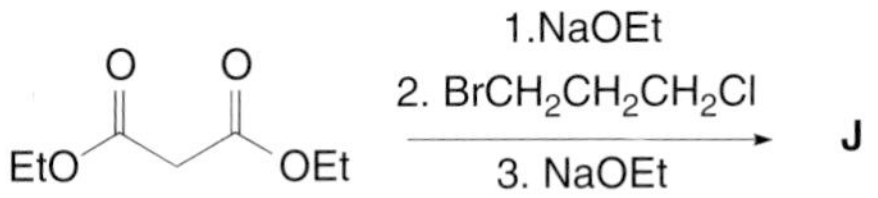

6

Alkenes and alkynes

Key point. Alkenes and alkynes are unsaturated hydrocarbons, which possess a C=C double bond and a C≡C triple bond, respectively. As (weak) π-bonds are more reactive than (strong) σ-bonds, alkenes and alkynes are more reactive than alkanes. The electron-rich double or triple bond can act as a nucleophile and most reactions of alkenes/alkynes involve *electrophilic addition reactions*. In these reactions, the π bond attacks an electrophile to generate a carbocation, which then reacts with a nucleophile. Overall, these reactions lead to the addition of two new substituents at the expense of the π bond.

6.1 Structure

Alkenes have a C=C double bond. The two carbon atoms in a double bond are sp^2 hybridised and the C=C bond contains one (strong) σ-bond and one (weaker) π-bond.

    Alkynes have a C≡C triple bond. The two carbon atoms in a triple bond are sp hybridised and the C≡C bond contains one σ-bond and two π-bonds.

Hybridisation is discussed in Section 1.5

Naming alkenes and alkynes is discussed in Section 2.4

The R groups represent alkyl groups, e.g. a methyl (CH_3) or ethyl (CH_3CH_2) group

Keynotes in Organic Chemistry, Second Edition. Andrew F. Parsons.
© 2014 John Wiley & Sons, Ltd. Published 2014 by John Wiley & Sons, Ltd.

The greater the number of π-bonds, the shorter and stronger the carbon–carbon bond becomes.

For a table of bond enthalpies see Appendix 1

	$HC\equiv CH$		$H_2C=CH_2$		H_3C-CH_3
CC bond strength (kJ mol^{-1})	837	>	636	>	368
CC bond length (Å)	1.20	<	1.33	<	1.54

Steric effects are introduced in Section 4.4

As there is no free rotation around a $C=C$ double bond, substituted alkenes can have E- and Z-stereoisomers (Section 3.3.1). For disubstituted alkenes, for example, Z-isomers are less stable than E-isomers because of steric interactions between the two bulky groups on the same side of the molecule.

The two R groups attached to the C=C bond can be the same or different

steric strain

Z-isomer
less stable

E-isomer
more stable

An alkene is stabilised by alkyl substituents on the double bond. When the stability of alkene isomers are compared it is found that the greater the number of alkyl substituents on the $C=C$ double bond, the more stable the alkene is. As an approximation, it is the number of alkyl substituents rather than their identities that governs the stability of an alkene. This can be explained by hyperconjugation which involves the donation of electrons from a filled $C-H$ σ-orbital to an empty $C=C$ π^*-orbital.

Hyperconjugation is introduced in Section 1.6.2

Overlap between these orbitals leads to stabilisation of the double bond. The more alkyl substituents on the double bond, the greater the number of C–H bonds which can interact, and so the more stable the alkene is.

The orbitals overlap side-on (Section 4.10)

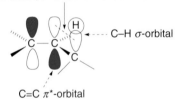

C–H σ-orbital

$C=C$ π^*-orbital

most stable   **least stable**

tetrasubstituted   trisubstituted   disubstituted   monosubstituted

It should be emphasised that both alkenes and alkynes have electron-rich π-bonds and so they react as nucleophiles.

Nucleophiles are defined in Section 4.2

6.2 Alkenes

6.2.1 Preparation

An important method for preparing alkenes involves the elimination (an E1 or E2 mechanism) of HX from halogenoalkanes, RX (Section 5.3.2). Alcohols (ROH) can also be converted into alkenes by activating the OH group (e.g. by protonation or conversion into a tosylate) to make this into a better leaving group (Section 5.2.2).

A reaction that leads to the loss of water is called a dehydration reaction

Tertiary amine oxides ($R_3N^+-O^-$) and xanthates (ROC(S)SR) can also undergo elimination to form alkenes on heating. These *Ei eliminations* (*elimination, intramolecular*) proceed via cyclic transition states, which require a synperiplanar arrangement of the leaving groups (i.e. the leaving groups lie in the same plane on the same side of the molecule). This can be compared to an E2 elimination, which requires an antiperiplanar arrangement of leaving groups (Section 5.3.2.1).

The use of hashed and wedged line notation is discussed in Section 3.3.2

The Cope elimination

Tertiary amine oxide   *5-membered transition state*

When both R groups are on the same side in the synperiplanar arrangement, the *Z*-isomer of the alkene is formed; when the R groups are on the opposite sides, the *E*-isomer of the alkene is formed

The Chugaev elimination

Xanthate

6-membered transition state

Alkenes can also be formed using the Wittig reaction (Section 8.3.4.3) and by reducing alkynes. *Z*-Alkenes are prepared on hydrogenation of alkynes in the presence of a Lindlar catalyst, and *E*-alkenes are formed from reaction of alkynes with Na/NH₃ (Section 6.3.2.4).

For assigning *Z*- and *E*-configurations, see Section 3.3.1.2

6.2.2 Reactions

Alkenes are nucleophiles and react with electrophiles in *electrophilic addition reactions*. These reactions lead to the introduction of two new substituents at the expense of the π-bond. (Some of these reactions were introduced in Section 5.2.3.)

For unsymmetrical alkenes, with different substituents at either end of the double bond, the electrophile adds regioselectively so as to form the more substituted (and therefore the more stable) carbocation.

Tertiary carbocations (R_3C^+) are more stable than secondary (R_2CH^+) or primary (RCH_2^+) carbocations (Section 4.3.1)

First step is the addition of an electrophile

The most stable (tertiary) carbocation is formed

For +I and −I effects, see Section 1.6.1

The more alkyl (R) groups on the double bond, the faster the rate of electrophilic addition. This is because the electron donating (+I) alkyl groups make the double bond more nucleophilic (e.g $Me_2C{=}CMe_2$ is much more reactive to electrophiles than $H_2C{=}CH_2$). Conversely, the introduction of electron withdrawing (−I, −M) substituents reduces the rate (e.g. $H_2C{=}CH{-}CH_3$ is much more reactive to electrophiles than $H_2C{=}CH{-}CF_3$).

6.2.2.1 Addition of hydrogen halides

The addition of HX (X = Cl, Br, I) to an alkene, to form halogenoalkanes, occurs in two steps. The first step involves addition of a proton (i.e. the electrophile) to the C=C bond to make the more stable intermediate carbocation. The second step involves nucleophilic attack by the halide anion. This produces a racemic halogenoalkane because the carbocation is planar and so can be attacked equally from either face. (These addition reactions are the reverse of halogenoalkane elimination reactions.)

A racemate is a 1:1 mixture of a pair of enantiomers; an S_N1 reaction also produces a racemate (via an intermediate carbocation), see Section 5.3.1.2

A tertiary carbocation is stabilised by three +I alkyl groups (Section 1.6.1)

The proton adds to the less substituted carbon atom

tertiary carbocation

secondary carbocation

The tertiary rather than the secondary carbocation is formed because this is more stable

The regioselective addition of HX to alkenes produces the more substituted halogenoalkane, which is called the Markovnikov (Markovnikoff) product. Markovnikov's rule states that: 'on addition of HX to an alkene, H attaches to the carbon with fewest alkyl groups and X attaches to the carbon with most alkyl groups.' This can be explained by the formation of the most stable intermediate carbocation.

Occasionally, the intermediate carbocations can also undergo structural rearrangements to form more stable carbocations. A hydrogen atom (with its pair of electrons) can migrate on to an adjacent carbon atom in a '*hydride shift*'. This will only occur if the resultant cation is more stable than the initial cation.

A regioselective reaction is a reaction that leads to the selective formation of one structural isomer over another

The name 1,2-hydride shift is sometimes used; 1,2- indicates that H$^-$ moves to an adjacent carbon

regioselective addition of a proton to give the secondary, rather than primary, carbocation

the hydride shift produces a more stable tertiary carbocation

Alkyl groups can also migrate to a carbocation centre in *Wagner-Meerwein rearrangements* (or *shifts*).

secondary carbocation   *tertiary* carbocation

The term 1,2-alkyl shift is sometimes used to describe the migration

Addition of HBr to alkenes in the presence of a peroxide

Whereas HBr usually adds to alkenes to give the Markovnikov product, in the presence of a peroxide (ROOR), HBr adds to give the alternative regioisomer. This is known as the anti-Markovnikov product. The change in regioselectivity occurs because the mechanism of the reaction changes from an ionic (polar) mechanism in the absence of peroxides, to a radical mechanism in the presence of peroxides. Initiation by peroxide leads to bromine radicals (or atoms) which add to the less hindered end of the alkene. This gives the more substituted radical, which in turn produces the less substituted halogenoalkane. (It should be remembered that radical stability follows the same order as cation stability – see Section 4.3.)

Notice the use of fishhook curly
arrows (Section 4.1)

RO⌒OR —heat or light→ RO• + •OR (Initiation step)

RO• + H⌒Br ⟶ RO—H + •Br

A strong O–H bond is formed at the expense of a weaker H–Br bond

The bromine atom adds to
the less hindered carbon
atom because this is more
accessible

secondary radical

The bromine atom is
regenerated and reacts with
another molecule of alkene
in a chain reaction

Curly arrows show the
formation of a tertiary
radical, which is more
stable than the
secondary radical

tertiary radical

**Anti-
Markovnikov
product**

6.2.2.2 Addition of bromine

In a stereospecific reaction
different stereoisomers of the
alkene produce different
stereoisomers of the product

The electrophilic addition of bromine produces a 1,2-dibromide (or vicinal dibromide). The addition is stereospecific because of the formation of an intermediate bromonium ion. This ensures that the bromine atoms add to opposite sides of the alkene in an *anti* addition.

Br⁻ can attack either carbon atom
of the bromonium ion; if the R
groups are the same, then attack is
equally likely at either carbon

Bromine reacts equally well
with both faces of the C=C
bond to give a racemic
bromonium ion

bromonium ion

The bromine molecule becomes
polarised as it approaches the alkene.
The bromine atom nearest the double
bond becomes electrophilic (as the
electrons in the Br–Br bond are repelled
away from the electron-rich double bond)

(cf. S_N2 reaction)

Overall *anti* addition as the two bromine atoms
end up on opposite sides of the planar alkene

1,2-dibromide

A racemic 1,2-dibromide is formed–the relative stereochemistry
of the bromonium ion and 1,2-dibromide is shown

The *anti* addition explains the formation of different diastereoisomers when using *E* or *Z* alkenes.

The relative stereochemistry of a diastereoisomer is the configuration of a diastereoisomer relative to another diastereoisomer. The absolute configuration of a diastereoisomer is the configuration of a single enantiomer

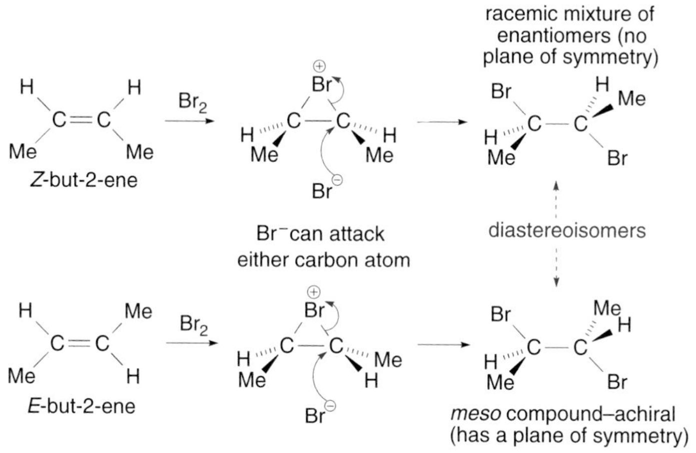

racemic mixture of enantiomers (no plane of symmetry)

Br⁻ can attack either carbon atom

diastereoisomers

Z-but-2-ene

meso compound–achiral (has a plane of symmetry)

E-but-2-ene

Meso compounds are discussed in Section 3.3.2.5

As Br⁻ can attack either carbon atom of the bromonium ion, reaction with *Z*-but-2-ene produces a 1:1 mixture of enantiomers (only the 2*R*,3*R* isomer is shown above). For *E*-but-2-ene, attack of Br⁻ at either carbon atom of the bromonium ion produces the same compound. This compound has a plane of symmetry and so is an achiral *meso* compound.

N-Bromosuccinimide or NBS (a stable, easily handled solid), rather than hazardous liquid bromine, is often used in these bromination reactions as this produces bromine (at a controlled rate) on reaction with HBr (which is usually present in trace amounts in the NBS).

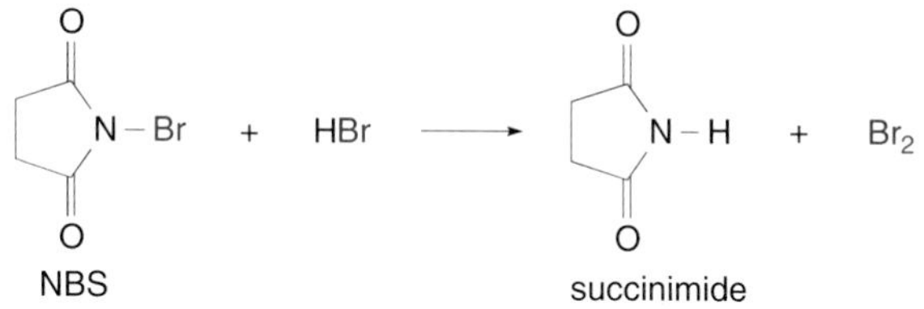

NBS   succinimide

Succinimide is an example of a cyclic imide; an imide has two acyl groups bonded to nitrogen (e.g. RCONHCOR)

6.2.2.3 Addition of bromine in the presence of water

The addition of bromine to an alkene in the presence of water can lead to the formation of a 1,2-bromo-alcohol (or bromohydrin) in addition to a 1,2-dibromide. This is because water can act as a nucleophile and compete with the bromide ion for ring-opening of the bromonium ion. Even though Br⁻ is a better nucleophile than H_2O, if H_2O is present in excess then the 1,2-bromo-alcohol will be formed.

S_N2 reactions are discussed in Section 5.3.1.1

An excess of water will produce the bromohydrin

$H_2\ddot{O}$

$+ Br^{\ominus}$

S_N2

As water is in excess, it is more likely to act as the base than Br⁻

$H_2\ddot{O}$

$H_3\overset{\oplus}{O} +$

Overall *anti* addition

bromohydrin

The bromohydrin is formed as a racemate – the relative stereochemistry is shown

The opening of the bromonium ion is often regioselective. The nucleophile usually attacks the more substituted carbon atom of the ring, because this carbon atom is more positively polarised. The reaction proceeds via a 'loose S_N2 transition state'.

Alkyl groups are electron-donating (+I) groups (Section 4.3.1)

then
$-H^{\oplus}$

Although less hindered, attack does not occur at this site

$H_2\ddot{O}$

The bromine begins to leave to produce a partial positive charge on the more substituted carbon. (The carbon substituents stabilise the $\delta+$ charge)

6.2.2.4 Addition of water (hydration): Markovnikov addition

Hydration of an alkene forms an alcohol. Dehydration of an alcohol forms an alkene (Section 6.2.1)

The addition of water to alkenes, to produce alcohols (ROH), requires the presence of (i) a strong acid or (ii) mercury(II) acetate (in an oxymercuration reaction). In both cases, the reactions involve the Markovnikov addition of water (i.e. the OH becomes attached to the more substituted carbon).

Acid catalysis (requires high temperatures)

Tertiary alcohols have the general formula R_3COH; secondary alcohols are R_2CHOH and primary alcohols are RCH_2OH

H_2O

unsymmetrical alkene

Most stable carbocation is formed

$H_2\ddot{O}$

$H_2\ddot{O}$

$H_3\overset{\oplus}{O} +$
(regenerated)

$R-\overset{\overset{\displaystyle R}{|}}{\underset{\underset{\displaystyle H-\overset{\overset{\displaystyle H}{|}}{\underset{\underset{\displaystyle H}{|}}{C}}-H}{|}}{C}}-OH$

overall addition of H_2O

tertiary alcohol

Oxymercuration

Tertiary carbocations (R_3C^+) are more stable than secondary carbocations (R_2CH^+), which are more stable than primary carbocations (RCH_2^+), see Section 4.3.1

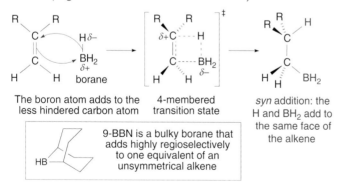

nucleophilic attack at the most substituted carbon atom with the greatest $\delta+$ character

mercurinium ion

unsymmetrical alkene

Reduction of the C–Hg bond to a C–H bond

tertiary alcohol

NaBH₄ (sodium borohydride)

Overall *anti* addition

Sodium borohydride ($NaBH_4$) is a reducing reagent, which is commonly used in synthesis (Section 8.3.3.1)

6.2.2.5   Addition of water (hydration): anti-Markovnikov addition

The anti-Markovnikov addition of water is achieved using borane (B_2H_6 which reacts as BH_3) in a hydroboration reaction. The reaction involves the *syn* addition of a boron–hydrogen bond to the alkene via a 4-membered transition state (i.e. the boron atom and the hydrogen atom add to the same face of the alkene). Hydroboration is a highly regioselective reaction and steric factors are important. The boron atom adds to the least hindered end of the alkene to give an organoborane (regioselectively), which is then oxidised to the alcohol. Bulkier boranes such as 9-BBN (9-borabicyclo[3.3.1] nonane) can enhance the regioselectivity of hydroboration.

In a syn addition, two atoms or groups add to the same side of the reactant

Hydroboration (regioselective and stereoselective)

The boron atom adds to the less hindered carbon atom   borane

4-membered transition state

syn addition: the H and BH₂ add to the same face of the alkene

9-BBN is a bulky borane that adds highly regioselectively to one equivalent of an unsymmetrical alkene

One equivalent is a condensed way of saying 'one stoichiometric equivalent', e.g. one mole of 9-BBN reacts with one mole of an alkene

Electronic factors also play a part. Boron is more electropositive than hydrogen, so the double bond will attack the boron atom to give a partial positive charge on one of the (alkene) carbon atoms in the 4-membered transition-state. The partial positive charge ($\delta+$) will reside on the more substituted carbon atom as this is more stable.

For the Pauling electronegativity scale, see Section 1.6.1

In the presence of two further alkene molecules, a trialkylborane (R_3B) is formed. Two further alkyl groups replace the two hydrogen atoms on boron (using the same mechanism as shown above).

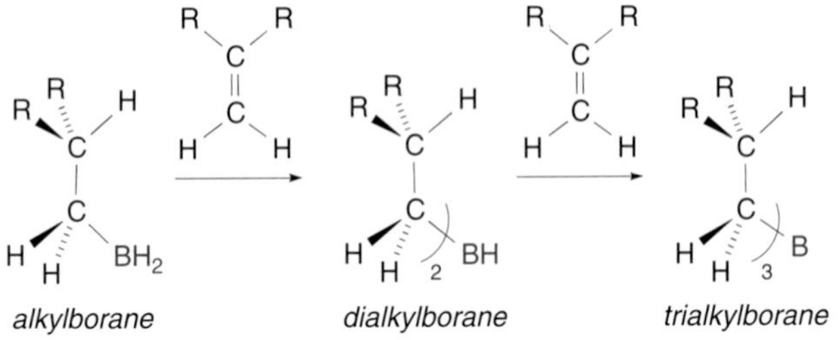

alkylborane   dialkylborane   trialkylborane

The trialkylborane is then oxidised using hydroperoxide (H_2O_2) in basic solution. This converts the three C−B bonds into three C−OH bonds.

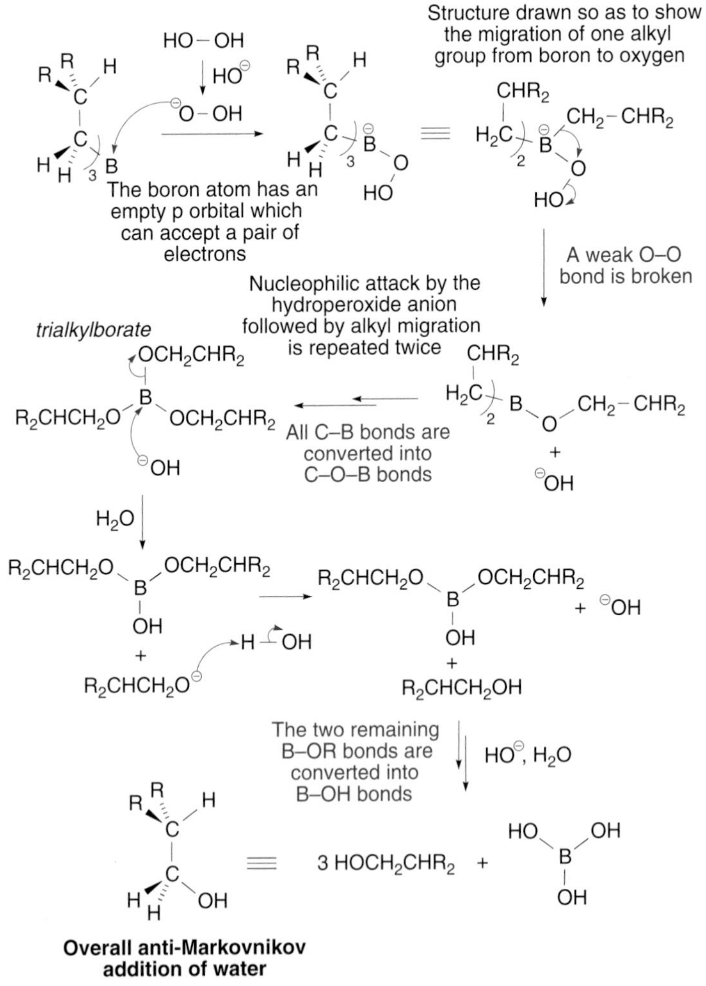

Structure drawn so as to show the migration of one alkyl group from boron to oxygen

The boron atom has an empty p orbital which can accept a pair of electrons

A weak O−O bond is broken

Nucleophilic attack by the hydroperoxide anion followed by alkyl migration is repeated twice

trialkylborate

All C−B bonds are converted into C−O−B bonds

The two remaining B−OR bonds are converted into B−OH bonds

B(OH)₃ is called boric acid

This reaction leads to a primary alcohol

Overall anti-Markovnikov addition of water

The migration of the alkyl groups from boron to oxygen occurs with retention of configuration.

The configuration of chiral centres is discussed in Section 3.3.2.2

The boron group is replaced by an hydroxyl group with the same stereochemistry

Two S_N2 reactions also lead to retention of configuration (Section 5.3.1.8)

6.2.2.6 Oxidation by peroxycarboxylic acids (RCO₃H) and hydrolysis of epoxides to give anti dihydroxylation

Reaction of alkenes with peroxycarboxylic acids (or peracids, RCO_3H) leads to the formation of epoxides in a concerted addition reaction, called epoxidation. The peroxycarboxylic acid donates an oxygen atom to the C=C double bond.

Epoxides can also be formed by S_N2 reactions (Section 5.3.1.8)

The addition is stereospecific; the *cis*-epoxide is formed from the *cis*-alkene. Epoxidation of a *trans*-alkene gives a *trans*-epoxide

The peroxycarboxylic acid reacts equally well with both faces of the alkene and so a racemic epoxide is formed (the structure for the *cis*-epoxide shows the relative stereochemistry)

Epoxides can be hydrolysed under acid- or base-catalysed conditions to form 1,2-diols (or glycols). These reactions lead to opening of the strained 3-membered (epoxide) ring. The formation and hydrolysis of an epoxide leads to the stereoselective *anti* addition of two OH groups to a C=C bond.

In a hydrolysis reaction, a reactant reacts with water

Base catalysis (aqueous hydroxide and heat)

Even though hydroxide is a poor nucleophile it will open the strained epoxide ring on heating

S_N2 reactions are discussed in Section 5.3.1.1

HO^- or H_2O can attack either carbon atom of the epoxide, or the protonated epoxide, respectively; if the R groups are the same, then attack is equally likely at either carbon

Acid catalysis (aqueous acid)

In *anti*-addition two substituents are added to opposite faces; this produces a single diastereoisomer (in a diastereoselective reaction)

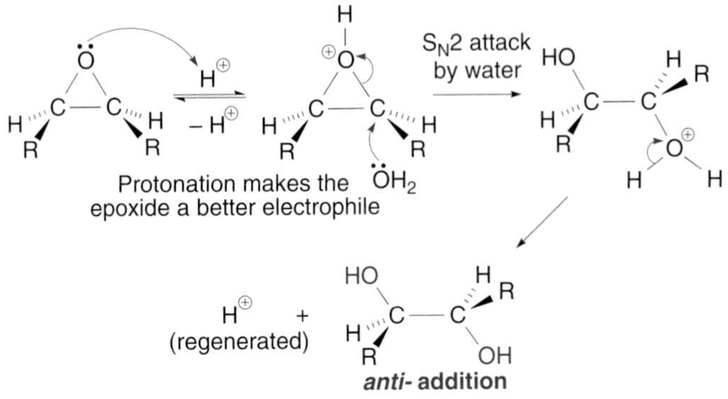

Protonation makes the $\overset{\cdot\cdot}{O}H_2$ epoxide a better electrophile

$H^{\oplus}$ + (regenerated)

***anti*- addition**

6.2.2.7 Syn *dihydroxylation and oxidative cleavage of 1,2-diols to form carbonyls*

A dihydroxylation reaction introduces two OH groups

Reaction of alkenes with potassium permanganate (KMnO₄) at low temperature, or osmium tetroxide (OsO₄), leads to the *syn* addition of two OH groups (i.e. the two OH groups add to the same face of the C=C bond).

Potassium permanganate (low temperature)

The dihydroxylation is stereospecific; a *cis*-alkene produces a *cis*-1,2-diol, a *trans*-alkene produces a *trans*-1,2-diol

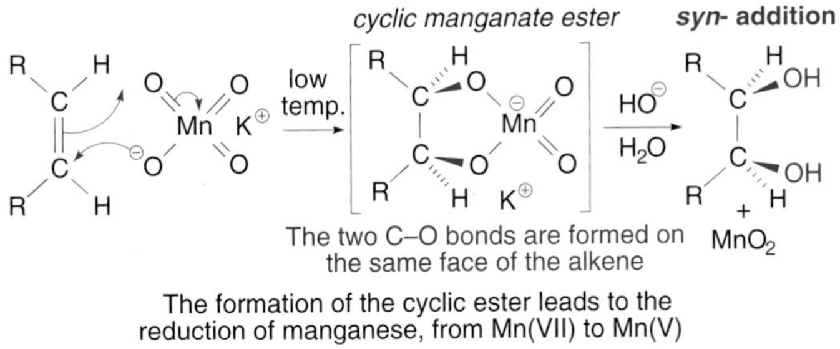

cyclic manganate ester   **syn- addition**

The two C–O bonds are formed on the same face of the alkene

The formation of the cyclic ester leads to the reduction of manganese, from Mn(VII) to Mn(V)

Osmium tetroxide

During the reaction, osmium is reduced from oxidation state VIII (in OsO₄) to VI (in H₂OsO₄).

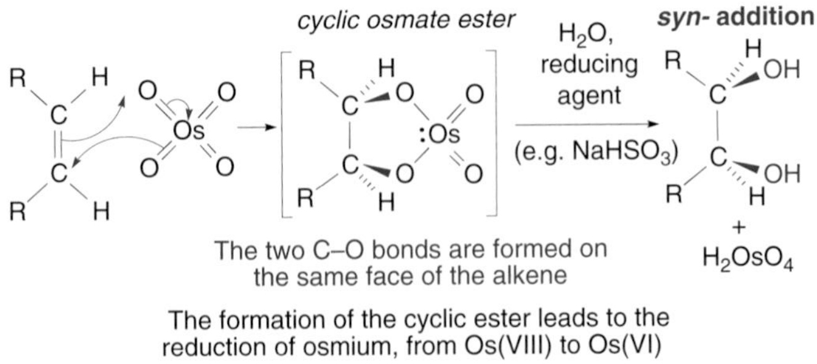

cyclic osmate ester   **syn- addition**

The two C–O bonds are formed on the same face of the alkene

The formation of the cyclic ester leads to the reduction of osmium, from Os(VIII) to Os(VI)

1,2-Diols, RCH(OH)CH(OH)R, can be oxidised further to form carbonyl compounds. This results in cleavage of a C–C bond. When potassium permanganate is reacted with alkenes at room temperature or above, the intermediate 1,2-diol can be further oxidised further; intermediate aldehydes (RCHO) are converted into carboxylic acids (RCO$_2$H).

Aldehydes can also be converted into carboxylic acids using CrO$_3$/ H$^+$ (Section 8.3.3.5)

disubstituted alkene   two carboxylic acids

1,2-Diols, RCH(OH)CH(OH)R, can also be oxidised to aldehydes and/or ketones by reaction with periodic acid (HIO$_4$).

1,2-diol   cyclic periodate ester   aldehyde

The byproduct, HIO$_3$, is called iodic acid; during the reaction iodine is reduced from oxidation state VII (in HIO$_4$) to V (in HIO$_3$).

6.2.2.8 Oxidative cleavage by reaction with ozone (O$_3$)

Reaction of alkenes with ozone at low temperature produces an intermediate molozonide (or primary ozonide), which rapidly rearranges to form an ozonide. This leads to cleavage of the C=C double bond.

As ozone has both a positive and a negative charge, it is called a dipolar reagent

This is known as a 1,3-dipolar addition (ozone is a 1,3-dipole)

molozonide (or primary ozonide)

The rearrangement leads to formation of two C–O bonds at the expense of a C–C bond and a very weak O–O bond

flip over aldehyde

ozonide

The intermediate ozonide can then be reduced to aldehydes (RCHO)/ketones (RCOR) or oxidised to carboxylic acids (RCO$_2$H).

Me$_2$S is called dimethyl sulfide (DMS); PPh$_3$ is called triphenylphosphine

ozonide

Aldehydes and carboxylic acids are also formed on oxidation of alcohols (Section 8.3.3.5)

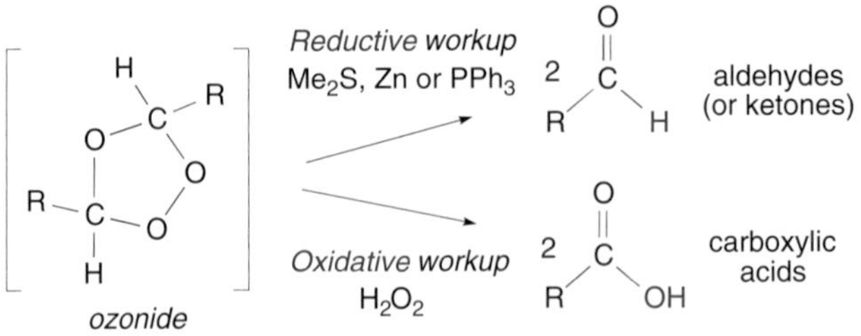

6.2.2.9   Reaction with carbenes

A carbene is a neutral, highly reactive molecule containing a divalent carbon with only 6 valence electrons. They are usually formed in one of two ways.

1. Reaction of trichloromethane (chloroform) and base. This is called an *α-elimination* reaction, as two groups (i.e. H and Cl) are eliminated from the *same* atom.

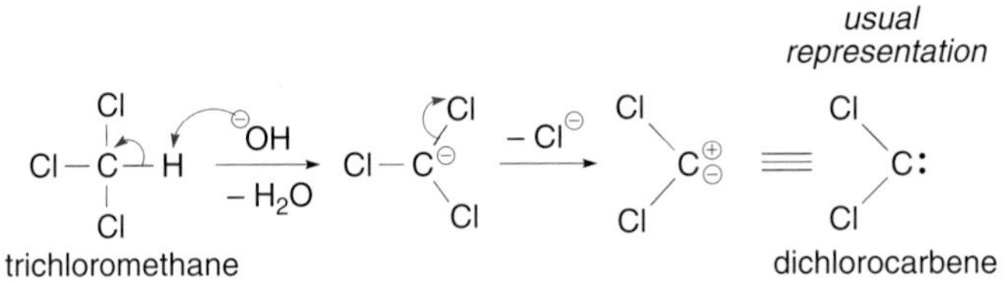

Loss of HX from RCH_2CH_2X, to form $RCH=CH_2$, is called a *β*-elimination; the H and X atoms are removed from adjacent atoms (Section 5.3.2)

trichloromethane   dichlorocarbene

2. Reaction of diiodomethane (CH_2I_2) and a zinc-copper alloy (in the Simmons-Smith Reaction).

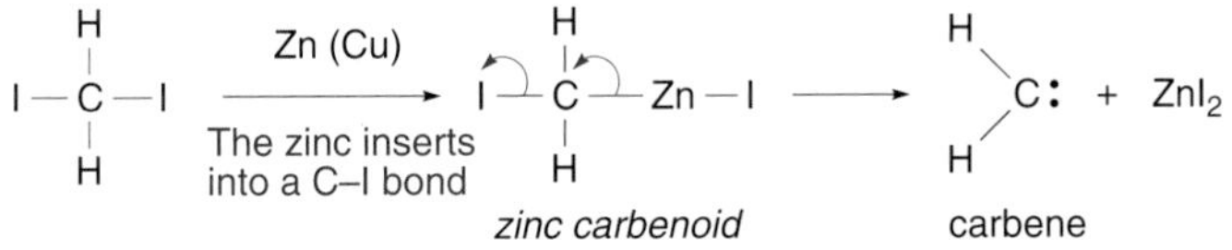

Although a free carbene is drawn here, the active reagent is the zinc carbenoid, which acts as a carbene equivalent

zinc carbenoid   carbene

As carbenes are electron-deficient, they can act as powerful electrophiles. At the same time, they can also be thought of as acting as nucleophiles because they contain an unshared pair of electrons.

For the conformation of cycloalkanes, including cyclopropane, see Section 3.2.3

Carbenes add to alkenes to form cyclopropanes. The mechanism of these insertion reactions depends on whether a singlet carbene (this contains a pair of reactive electrons in the same orbital) or a triplet carbene (this contains two unpaired and reactive electrons in different orbitals) adds to the double bond.

Singlet carbene (concerted mechanism – stereospecific syn-addition)

A racemic cyclopropane is formed as the carbene reacts equally with the top and bottom face of the C=C bond

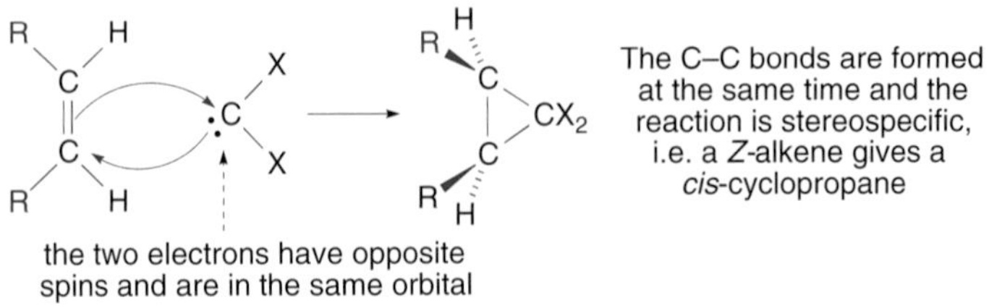

the two electrons have opposite spins and are in the same orbital

The C–C bonds are formed at the same time and the reaction is stereospecific, i.e. a *Z*-alkene gives a *cis*-cyclopropane

Triplet carbene (stepwise mechanism – non-stereospecific)

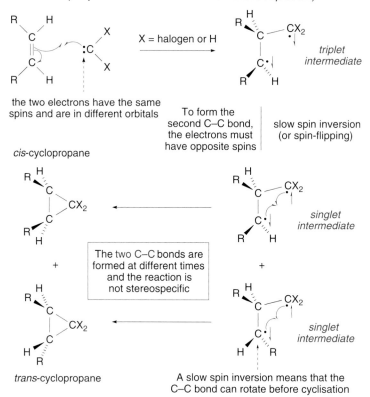

Notice the use of single-headed (or fishhook) curly arrows to represent the movement of single electrons (Section 4.1)

the two electrons have the same spins and are in different orbitals

cis-cyclopropane

To form the second C–C bond, the electrons must have opposite spins

slow spin inversion (or spin-flipping)

The two C–C bonds are formed at different times and the reaction is not stereospecific

A *cis*-cyclopropane has the groups of highest priority on the same side of the ring; for assigning priorities to groups, see Section 3.3.2.2

trans-cyclopropane

A slow spin inversion means that the C–C bond can rotate before cyclisation

6.2.2.10 Addition of hydrogen (reduction)

The alkene double bond can be hydrogenated by hydrogen in the presence of a platinum or palladium catalyst. The hydrogenation occurs with *syn* stereochemistry as the two hydrogen atoms are added to the same face of the alkene on the surface of the catalyst.

Catalysts are discussed in Section 4.9.2.1

metal catalyst surface

6.2.2.11 Reaction with dienes

Alkenes bearing an electron-withdrawing substituent(s) can add to electron-rich conjugated dienes in the *Diels-Alder cycloaddition reaction*. This is a concerted reaction leading to the formation of two new C–C bonds in one step. It is an important method for making rings including cyclohexenes.

The Diels-Alder reaction is an example of a pericyclic reaction (Section 4.6.3)

−I and −M groups are introduced in Section 1.6

1,3-diene   The alkene is called the dienophile   cyclic transition state   A substituted cyclohexene

This is a *pericyclic* reaction, which involves a concerted redistribution of bonding electrons (i.e. a pericyclic reaction is a concerted reaction that involves a flow of electrons). The preference for electron-rich dienes and electron-deficient dienophiles arises from orbital interactions; this combination gives a good overlap of orbitals in the transition-state.

Orbital overlap and energy is discussed in Section 4.10

The s-*cis* (or cisoid) conformation of the diene is required and the reaction is stereospecific (i.e. only *syn*-addition). Groups that are *cis* in the (alkene) dienophile will therefore be *cis* in the product. Conversely, groups that are *trans* in the dienophile will be *trans* in the product.

A racemic product is formed as the 1,3-diene reacts equally with the top and bottom face of the C=C bond

s-*trans*   s-*cis*   *cis*-alkene   *cis*-product

6.3 Alkynes

6.3.1 Preparation

Alkynes can be formed from 1,2-dibromoalkanes (or vicinal dihalides) by elimination of two molecules of HX using a strong base (e.g. sodium amide, NaNH$_2$).

A disubstituted alkyne is sometimes called an internal alkyne. A monosubstituted alkyne (RC≡CH) is sometimes called a terminal alkyne

disubstituted alkene   1,2-dibromoalkane   disubstituted alkyne

6.3.2 Reactions

Alkynes, like alkenes, act as nucleophiles and react with electrophiles in electrophilic addition reactions. The electrophiles generally add in the same way as they add to alkenes.

6.3.2.1 Addition of hydrogen halides (HX)

The addition of HX occurs so as to give the most substituted halogenoalkane (RX) following Markovnikov's rule (Section 6.2.2.1). The use of one or two equivalents of HX can be used to form vinyl halides (halogenoalkenes) or dihaloalkanes, respectively.

The R group is electron donating and stabilises the vinylic carbocation by a +I effect (Section 4.3.1)

The addition proceeds via the most stable (or substituted) vinylic carbocation

Vinylic carbocations (e.g. $RC^+=CH_2$) are generally less stable than alkyl carbocations (e.g. RCH^+CH_3) as there are fewer +I alkyl groups to stabilise the positive charge. As a consequence, alkynes (which give vinylic carbocations) generally react more slowly than alkenes (which give alkyl carbocations) in electrophilic addition reactions.

6.3.2.2 Addition of water (hydration): Markonikov addition

Alkynes will react with water in the presence of a mercury(II) catalyst, such as mercury(II) acetate. Water adds in a Markovnikov addition to form an enol (e.g. $RCH(OH)=CH_2$), which tautomerises to give a ketone, e.g. $RCOCH_3$ (see Section 8.4.1).

Compare the mechanism of formation of the 3-membered ring (a bridged mercurinium ion), with the formation a bromonium ion, Section 6.2.2.2

Regioselective ring opening is explained by formation of a partial positive charge ($\delta+$) on the more substituted carbon

A catalyst is not consumed by the reaction (Section 4.9.2.1)

6.3.2.3 Addition of water (hydration): anti-Markonikov addition

The hydroboration of alkynes using borane (BH_3) produces an intermediate vinylic borane (e.g. $RCH{=}CHBH_2$), which then reacts with BH_3 in a second hydroboration; the two boron groups add to the less hindered end of the triple bond. However, when a bulky dialkylborane (HBR_2) is used, steric hindrance prevents a second hydroboration and oxidation of the intermediate vinylic borane produces an enol (e.g. $RCH{=}CHOH$), which tautomerises to an aldehyde (e.g. RCH_2CHO). (For a related mechanism see Section 6.2.2.5.)

Steric hindrance is introduced in Section 4.4

Keto-enol tautomerism is discussed in Section 8.4.1

This results in the anti-Markovnikov addition of water to the triple bond.

6.3.2.4 Addition of hydrogen (reduction)

Alkynes can be stereoselectively reduced to give Z- or E-alkenes using H_2/Lindlar catalyst or Na/NH_3 (at low temperature), respectively.

Z-alkene

Assigning *E* and *Z* alkenes is described in Section 3.3.1.2

The metal catalyst ensures that the two hydrogen atoms add to the same face of the alkene in a *syn*-addition.

Catalysts are discussed in Section 4.9.2.1

The Lindlar catalyst [$Pd/CaCO_3/PbO$] is a poisoned palladium catalyst, which ensures that the reduction stops at the alkene (and does not go on to give an alkane). Reduction to the alkane requires a Pd/C catalyst (Section 6.2.2.10).

E-alkene

Adding a single electron to a neutral molecule forms a radical anion

Steric strain is introduced in Section 3.2.2

The two hydrogen atoms add to the opposite faces of the alkene (i.e. *anti*-addition) using sodium in liquid ammonia. This 'dissolving metal reduction' produces a solvated electron, which adds to the alkyne to produce a radical anion (bearing a negative charge and an unpaired electron). The bulky R groups in the radical anion, vinyl radical and subsequently in the vinyl anion lie as far apart as possible to avoid steric strain.

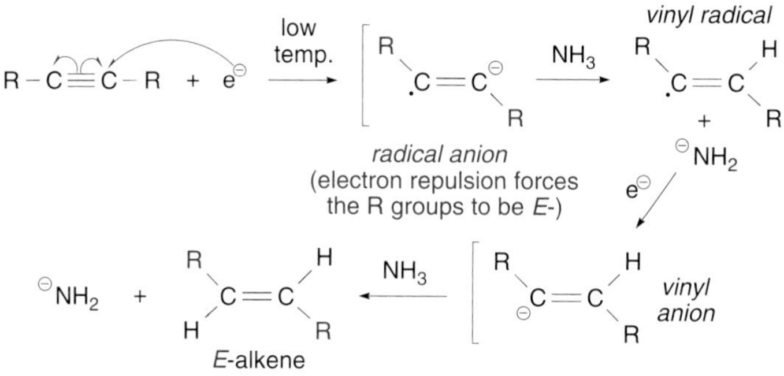

An *E*-alkene has the highest priority groups on the opposite sides of the C=C bond (Section 3.3.1.2)

6.3.2.5 Deprotonation: formation of alkynyl anions

Terminal alkynes (RC≡CH) contain an acidic proton, which can be deprotonated by sodium amide (NaNH$_2$). The negative charge of the alkynyl (or acetylide) anion (e.g. RC≡C$^-$) resides in an sp orbital. This is more stable than a vinyl anion, produced on deprotonation of an alkene (e.g. RCH=CH$^-$), because these are sp^2 hybridised. The greater the 's' character, the more closely the anion is held to the positively charged nucleus (which stabilises it).

The alkynyl anion is a strong nucleophile. Reaction with primary halogenoalkanes (RCH$_2$X) will lead to alkylation (by an S$_N$2 mechanism) and the introduction of an alkyl group (R^1 below) on the terminal carbon atom of the alkyne.

The acidity of terminal alkynes is introduced in Sections 1.7.4 and 1.7.6

S$_N$2 reactions are introduced in Section 5.3.1.1

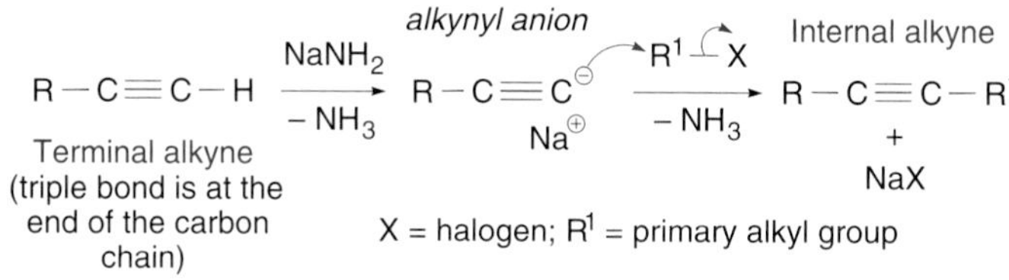

This type of reaction is useful in synthesis as it forms a new C–C bond – it can be used to make larger products from smaller reactants

Worked example

The following questions relate to the synthesis of (±)-diol **4** from alkyne **1**, shown below. [(±) indicates a racemic compound.]

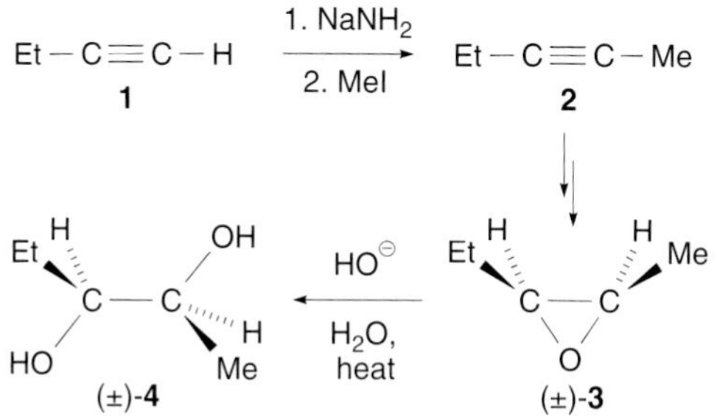

Hint: NaNH$_2$ is a strong base

Hint: Compound **3** is an epoxide, which can be prepared from an alkene

Hint: Consider a concerted nucleophilic substitution reaction

Hint: Is a single enantiomer and/or a single diastereoisomer formed?

Hint: Identify the relationship between (±)-**4** and (±)-**5**

(a) Give a reaction mechanism to explain the conversion of alkyne **1** into alkyne **2**.

(b) Give the reagents needed to convert alkyne **2** into compound (±)-**3**, in two steps.

(c) Give a reaction mechanism to explain the conversion of compound (±)-**3** into compound (±)-**4**.

(d) Is the formation of compound (±)-**4** an example of an enantioselective synthesis and/or a diastereoselective synthesis?

(e) Give the reagents needed to convert alkyne **2** into diol (±)-**5**, in three steps.

(±)-**5**

Formation and reaction of alkynyl anions is discussed in Section 6.3.2.5

Answer

(a)

For reduction of alkynes and epoxidation of alkenes see Sections 6.3.2.4 and 6.2.2.6, respectively

(b) 1. H$_2$, Lindlar catalyst. 2. RCO$_3$H

(c)

Base catalysed hydrolysis of an epoxide is discussed in Section 6.2.2.6

(±)-**3**   (±)-**4**

Enantiomers and diastereoisomers are discussed in Sections 3.3.2.1 and 3.3.2.4

For the stereoselective reduction of alkynes see Section 6.3.2.4

(d) The synthesis leads to an equal mixture of enantiomers of diol **4** (a racemate is formed) and so the reaction is not enantioselective. This is an example of a diastereoselective synthesis. A single diastereoisomer of **4** is produced – opening of the epoxide ring leads to overall *anti-* addition of the two OH groups.

(e) 1. Na, NH$_3$. 2. RCO$_3$H. 3. HO$^-$, H$_2$O

Problems

Section 6.2.2.1

1. Give mechanisms for the reaction of HBr with pent-1-ene in (a) the presence of a peroxide (ROOR) and (b) in the absence of a peroxide. Explain how your mechanisms account for the regioselectivity of the reactions.

Section 6.2.2.2

2. Give a mechanism for the reaction of Br$_2$ with (Z)-hex-3-ene, and explain how the mechanism accounts for the stereospecificity of the reaction.

3. Suggest suitable reagents to accomplish the following transformations. Note that the *relative* rather than *absolute stereochemistry* of the products is shown (i.e. even though just one enantiomer of the product is drawn, the product is racemic because the alkene starting material is achiral).

Sections 6.2.2.5–8 and 6.2.2.10

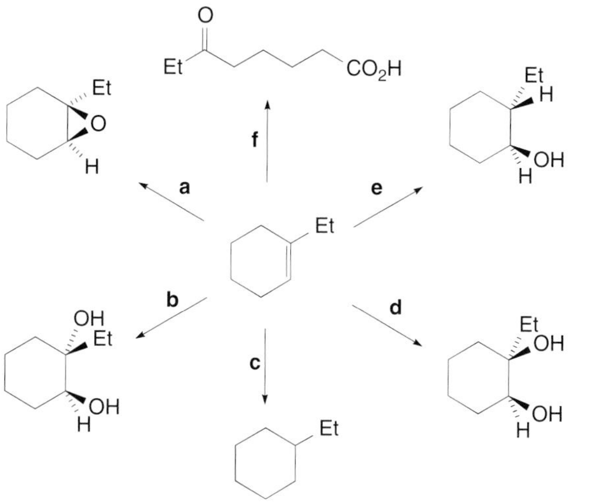

4. Suggest syntheses of compounds **A**, **B** and **C** starting from the alkyne CH₃CH₂CH₂C≡CH (pent-1-yne).

Sections 6.3.2.2, 6.3.2.4 and 6.3.2.5

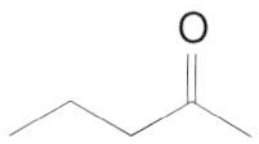

5. Give the diene and dienophile that would react in a Diels-Alder reaction to give each of the products **D-F**.

Section 6.2.2.11

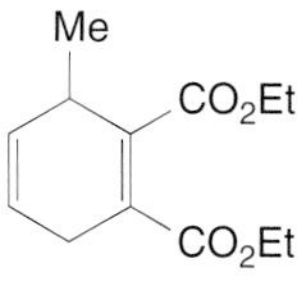

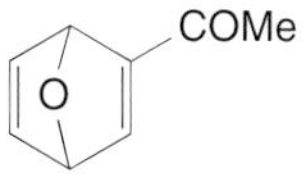

6. Explain why reaction of buta-1,3-diene ($H_2C{=}CH{-}CH{=}CH_2$) with one equivalent of HCl produces both 3-chlorobut-1-ene (~80%) and 1-chlorobut-2-ene (~20%).

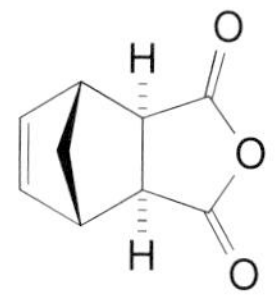

Section 6.2.2.1

7. The following questions relate to the synthesis of compound **I**, shown below.

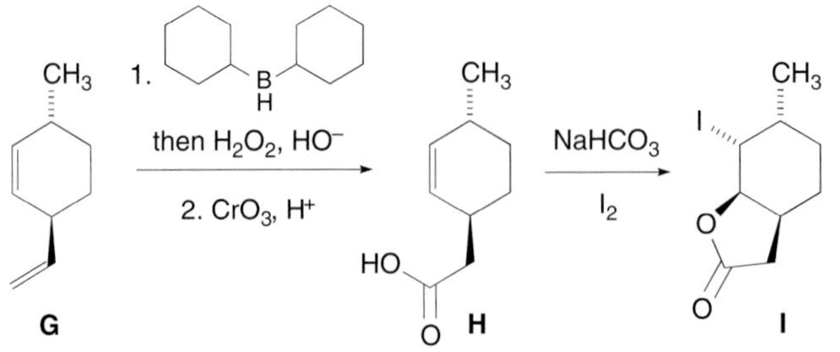

Section 6.2.2.5

(a) Explain the two-step conversion of **G** into **H** and discuss any reaction selectivity.

Section 6.2.2.2

(b) Assuming that I_2 reacts with a C=C bond in the same way as Br_2, propose a reaction mechanism to explain the stereoselective formation of compound **I**.

7

Benzenes

Key point. Benzene is an aromatic compound because the six π electrons are delocalised over the planar 6-membered ring. The delocalisation of electrons results in an increase in stability and benzene is therefore less reactive than alkenes or alkynes. Benzene generally undergoes *electrophilic substitution reactions*, in which a hydrogen atom is substituted for an electrophile. The electron-rich benzene ring attacks an electrophile to form a carbocation, which rapidly loses a proton so as to regenerate the aromatic ring. Electron-donating (+I, +M) substituents (on the benzene ring) make the ring more reactive towards further electrophilic substitution and direct the electrophile to the *ortho-/para-* positions. In contrast, electron-withdrawing (−I, −M) substituents (on the benzene ring) make the ring less reactive towards further electrophilic substitution and direct the electrophile to the *meta-* position.

7.1 Structure

- Benzene (C_6H_6) has six sp^2 carbon atoms and is cyclic, conjugated and planar. It is symmetrical and all C−C−C bond angles are 120°. The six C−C bonds are all 1.39 Å long, which is in between the normal values for a C−C and C=C bond.

  Hybridisation is discussed in Section 1.5

- Benzene has six π electrons, which are delocalised around the ring and a circle in the centre of a 6-membered ring can represent this. However, it is generally

  Drawing a benzene ring is introduced in Section 2.5

Keynotes in Organic Chemistry, Second Edition. Andrew F. Parsons.
© 2014 John Wiley & Sons, Ltd. Published 2014 by John Wiley & Sons, Ltd.

shown as a ring with three C=C bonds, because it is easier to draw reaction mechanisms using this representation.

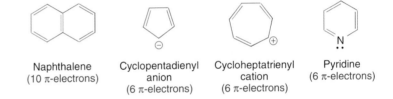

The six p-orbitals overlap to
give 'rings' of electron density
above and below the six-carbon ring

Circle denotes
delocalisation
of electrons

Shown as a
single resonance
form

Naming derivatives of benzene is
discussed in Section 2.4

As benzene has six π-electrons, it obeys Huckel's rule and is *aromatic*. Huckel's rule states that only cyclic planar molecules with $4n + 2$ π-electrons can be aromatic: for benzene $n = 1$. (Systems with $4n$ π-electrons are described as *anti-aromatic*.)

Pyridine is a weak base,
Section 1.7.5

Aromatic compounds can be monocyclic or polycyclic, neutral or charged. Atoms other than carbon can also be part of the ring and for pyridine, the lone pair of electrons on nitrogen is not part of the π-electron system (Section 1.7.5).

For electrophilic substitution of
naphthalene, see Section 7.9

Examples (each double bond represents two π-electrons)

For electrophilic substitution of
pyridine, see Section 7.10

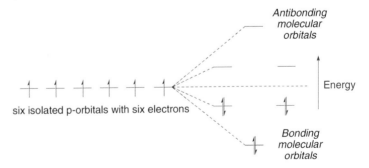

Naphthalene
(10 π-electrons)

Cyclopentadienyl
anion
(6 π-electrons)

Cycloheptatrienyl
cation
(6 π-electrons)

Pyridine
(6 π-electrons)

For benzene, molecular orbital theory shows that the six p-orbitals combine to give six molecular orbitals. The three lower-energy molecular orbitals are bonding molecular orbitals, and these are completely filled by the six electrons (which are spin-paired). There are no electrons in the (higher-energy) antibonding orbitals, and hence benzene has a closed bonding shell of delocalised π-electrons.

Orbital overlap and energy is
introduced in Section 4.10

Antibonding
molecular
orbitals

Energy

six isolated p-orbitals with six electrons

Bonding
molecular
orbitals

The delocalisation of electrons in aromatic compounds gives rise to characteristic chemical shift values in the ^{1}H NMR spectra (Section 10.6.1).

7.2 Reactions

The delocalisation of electrons means that benzene is unusually stable. It has a heat of hydrogenation which is approximately $150\,kJ\,mol^{-1}$ less than that would be expected for a cyclic conjugated triene. This is called the *resonance energy*.

A conjugated compound contains alternating single and multiple bonds (e.g. $H_2C{=}CH{-}CH{=}CH_2$)

The unusual stability of benzene (and other aromatic molecules) means that it undergoes *substitution* rather than addition reactions (cf. alkenes/alkynes). This is because substitution reactions lead to products that retain the stable aromatic ring.

Because benzene contains contains six π-electrons, it can act as a nucleophile and react with electrophiles in *electrophilic substitution reactions*. These reactions involve the substitution of a hydrogen atom on the benzene ring for an electrophile. The initial electrophilic attack on the benzene ring leads to a positively charged intermediate (called a *Wheland intermediate*), which is readily deprotonated.

Substitution reactions are introduced in Section 4.6.1.3

nucleophile

substituted benzene

The slowest step is called the rate-determining step (Section 4.9.2.1)

For an introduction to resonance, see Section 1.6.3

7.2.1 Halogenation

Reaction of benzene (C_6H_6) with bromine or chlorine in the presence of a Lewis acid catalyst (such as $FeBr_3$, $FeCl_3$ or $AlCl_3$) leads to bromobenzene (C_6H_5Br) or chlorobenzene (C_6H_5Cl), respectively. The Lewis acid, which does not have a full outer electron shell, can form a complex with bromine or chlorine. This polarises the halogen-halogen bond (making the halogen more electrophilic), and attack occurs at the positive end of the complex.

The Lewis acid catalyst is sometimes called a halogen carrier

Lewis acids are introduced in Section 1.7.3

The singly bonded halogen is electrophilic because electron density in the X–X bond moves to the positively charged halogen (which is a –I group, Section 1.6.1)

X = Br or Cl

The Lewis acid can accept a lone pair of electrons from the halogen

FeX_3 + HX +
(regenerated)

bromobenzene or chlorobenzene

For reaction of bromine with an alkene, see Section 6.2.2.2

No reaction occurs in the absence of a Lewis acid. This contrasts with the halogenation of alkenes/alkynes, which does not require activation by a Lewis acid. This is because alkenes/alkynes are not aromatic, and hence they are more reactive nucleophiles than benzene.

7.2.2 Nitration

Benzene (C_6H_6) can be nitrated, using a mixture of concentrated nitric acid (HNO_3) and sulfuric acid (H_2SO_4), to form nitrobenzene ($C_6H_5NO_2$). These acids react to form an intermediate nitronium ion ($^+NO_2$), which acts as the electrophile.

Sulfuric acid is a stronger acid than nitric acid

Water is a good leaving group (Section 5.3.1.4)

nitronium ion

It is likely that water, or $HOSO_3^-$, acts as a base to remove the proton in the nonaromatic carbocation

nonaromatic carbocation   *nitrobenzene*

7.2.3 Sulfonation

Reaction of benzene (C_6H_6) with fuming sulfuric acid or oleum (a mixture of sulfuric acid and sulfur trioxide, SO_3) leads to benzenesulfonic acid ($C_6H_5SO_3H$). The sulfonation is reversible and this makes it a useful tool in organic synthesis for blocking certain positions on a benzene ring (see Section 7.8). Sulfonation is favoured using strong sulfuric acid, but desulfonation is favoured in hot, dilute aqueous acid.

electrophile

It is likely that $HOSO_3^-$ acts as a base to remove the proton in the nonaromatic carbocation

nonaromatic carbocation   *benzenesulfonic acid*

Sulfur trioxide (rather than protonated sulfur trioxide, $HOSO_2^+$) could also act as the electrophile in these reactions.

The sulfonate ion can then be protonated

Oxygen is more electronegative than sulfur

$HOSO_2^+$ is a stronger electrophile than SO_3 (Section 4.2.1.2)

7.2.4   Alkylation: The Friedel-Crafts alkylation

Reaction of benzene (C_6H_6) with bromoalkanes (RBr) or chloroalkanes (RCl) in the presence of a Lewis acid catalyst (such as $FeBr_3$, $FeCl_3$ or $AlCl_3$) leads to alkylbenzenes (C_6H_5R). (Aryl or vinyl halides, ArX or R_2C=CHX, do not react.) The Lewis acid increases the electrophilicity of the carbon atom attached to the halogen. For *primary* haloalkanes (RCH_2X), a coordination complex is formed.

Primary (RCH_2X), secondary (R_2CHX) and tertiary (R_3CX) halogenoalkanes are introduced in Section 2.1

primary haloalkane   coordination complex

In the coordination complex, electron density in the C–X bond moves to the positively charged halogen (which is a –I group, Section 1.6.1)

FeX₃   +   HX   +   alkylbenzene
(regenerated)

For *secondary* (R_2CHX) and particularly *tertiary* (R_3CX) haloalkanes, this can lead to the formation of a carbocation (R_2CH^+ or R_3C^+), which then reacts with the electron-rich benzene ring.

Tertiary carbocations are more stable than secondary carbocations, which are more stable than primary carbocations (Section 4.3.1)

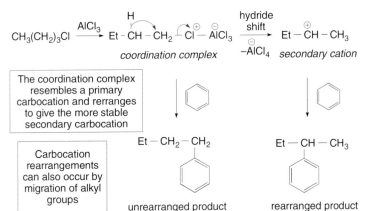

Intermediate coordination complexes (e.g. RCH_2X^+–Fe^-X_3) or secondary carbocations (R_2CH^+) may rearrange, to give more stable secondary or tertiary carbocations, before carbon–carbon bond formation can take place (see Section 6.2.2.1). In general, the higher the reaction temperature the greater the amount of rearranged product.

Et is the abbreviation for an ethyl group, CH_3CH_2

The movement of H^- to an adjacent carbon is called a 1,2-hydride shift (Section 6.2.2.1)

7.2.5 Acylation: The Friedel-Crafts acylation

Reaction of benzene (C_6H_6) with acyl chlorides (acid chlorides, RCOCl) in the presence of a Lewis acid (such as $FeCl_3$ or $AlCl_3$) leads to acylbenzenes (C_6H_5COR). The Lewis acid increases the electrophilicity of the carbon atom attached to the chlorine. This leads to the formation of an acylium ion, which reacts with the electron-rich benzene ring. Following acylation of the benzene ring, the acylbenzene forms a coordination complex with the Lewis acid. Treatment with water, at the end of the reaction, releases the acylbenzene.

One equivalent of the Lewis acid is required – it forms a coordination complex with the product and so does not act as a catalyst

acyl chloride · coordination complex · acylium ion

acylbenzene · coordination complex

Treatment with water at the end of the reaction is often described as an 'aqueous workup'

Unlike carbocations, the intermediate acylium ion does not rearrange and is attacked by the benzene ring to give exclusively the unrearranged product.

The Gatterman-Koch reaction (which uses CO, HCl and AlCl$_3$) can be used to form an aromatic aldehyde (e.g. benzaldehyde, PhCHO), rather than a ketone.

For non-systematic names of substituted benzenes, see Section 2.4

Sometimes copper(I) chloride, CuCl, is added to the reaction mixture

benzaldehyde

7.3 Reactivity of substituted benzenes

The introduction of substituents on the benzene ring affects both the reactivity of the benzene ring and also the regioselectivity of the reaction (i.e. the position in which the new group is introduced on the benzene ring).

Regioselectivity is introduced in Section 4.8

7.3.1 Reactivity of benzene rings: Activating and deactivating substituents

Inductive and mesomeric effects are discussed in Sections 1.6.1 and 1.6.3, respectively

The mechanism of electrophilic substitution of benzene is discussed in Section 7.2

- Substituents that *donate* electron density towards the benzene ring are known as the *activating groups*. These groups, which have positive inductive (+I) and/or mesomeric effects (+M), make the substituted benzene ring more reactive to electrophilic substitution than benzene itself. This is because the activating group can help to stabilise the carbocation intermediate (by donating electrons).
- Substituents that *withdraw* electron density away from the benzene ring are known as the *deactivating groups*. These groups, which have negative inductive (−I) and/or mesomeric effects (−M), make the substituted benzene ring less reactive to electrophilic substitution than benzene itself. This is because the deactivating group can destabilise the carbocation intermediate (by withdrawing electrons).

EDG (+I, +M)   More reactive than benzene   EDG   E   H   EDG stabilises the carbocation

EDG = electron donating group

EWG (−I, −M)   Less reactive than benzene   EWG   E   H   EWG destabilises the carbocation

EWG = electron withdrawing group

As nitrogen is less electronegative than oxygen, the NH$_2$ group has a stronger +M effect than the OH group

- The stronger the +M and/or +I effect, the more activating the group and the more reactive the benzene ring is to electrophilic attack. Note that positive mesomeric effects are generally stronger than positive inductive effects (see Section 1.6).
- The greater the −M and/or −I effect, the more deactivating the group and the less reactive the benzene ring is to electrophilic attack.

The SO$_3$H and NO$_2$ groups are most electron withdrawing – they exert a strong −I effect as they contain 3 or 4 electronegative atoms

Activating groups		Deactivating groups	
NHR, NH$_2$	(+M, −I)	Cl, Br, I	(+M, −I)
OR, OH	(+M, −I)	CHO, COR	(−M, −I)
NHCOR	(+M, −I)	CO$_2$H, CO$_2$R	(−M, −I)
Aryl (Ar)	(+M, +I)	SO$_3$H	(−M, −I)
Alkyl (R)	(+I)	NO$_2$	(−M, −I)

Most reactive ↑ (Activating groups)

Least reactive ↓ (Deactivating groups)

7.3.2 Orientation of reactions

Electrophilic substitution can occur at the *ortho-* (2-/6-), *meta-* (3-/5-) or *para-* (4-) positions of the benzene ring (see Section 2.4). The inductive and/or mesomeric effects of the existing substituent determine which position the new substituent is introduced on the ring. On introducing a new substituent, the formation of the carbocation intermediate is the rate-determining step. If the existing substituent can stabilise the carbocation, then this will lower the activation energy, leading to attack at this position.

The slowest step is the rate-determining step (Section 4.9.2.1)

Substituents are classified as (i) *ortho-/para-*directing activators; (ii) *ortho-/para-*directing deactivators; or (iii) *meta-*directing deactivators.

ortho-/para-activators	ortho-/para-deactivators	meta-deactivators
NHR, NH$_2$ OR, OH NHCOR Aryl (Ar) Alkyl (R)	Cl, Br, I	CHO, COR CO$_2$H, CO$_2$R SO$_3$H NO$_2$

ortho-, *meta-* and *para-*positions are introduced in Section 2.4

The relative ease of attack (by an electrophile) at the *ortho-/meta-/para-*positions can be determined experimentally from *partial rate factors*. These compare the rate of attack at one position in the mono-substituted benzene with the rate of attack at one position in benzene (under the same conditions). The higher the partial rate factor at a given position, the faster the rate of electrophilic substitution.

7.3.2.1 Ortho-/para- directing activators

Electron-donating +I and/or +M groups (EDG), which make the ring more nucleophilic than benzene, will stabilise the intermediate carbocation most effectively when new substituents are introduced at the *ortho-* or *para-*positions. For the *meta-* isomer, the positive charge in the carbocation intermediate does not reside adjacent to the EDG in any of the resonance forms.

Resonance forms are introduced in Section 1.6.3

EDG = electron donating group

ortho-

para-

meta-

Most stable

Least stable

The EDG can stabilise both of these cations more effectively than the cation derived from *meta-* attack

For stabilisation of carbocations by resonance, see Section 1.6.3

An additional resonance structure can be drawn for intermediate carbocations bearing a +M group, but not a +I group (at the 2- or 4-positions).

Examples

For an NH$_2$ or OH group, the +M
effect is stronger than the –I effect

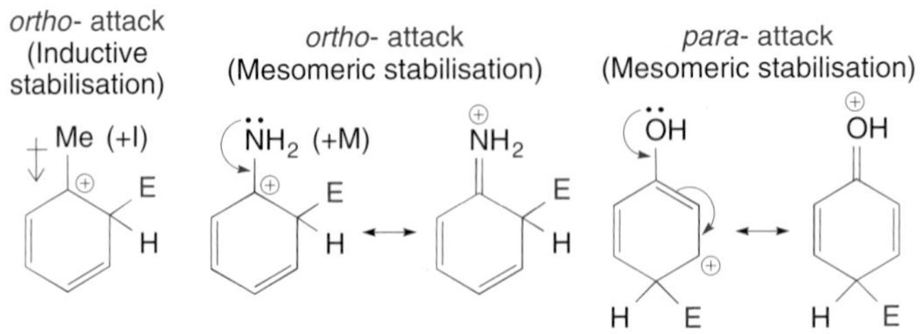

ortho- attack
(Inductive
stabilisation)

ortho- attack
(Mesomeric stabilisation)

para- attack
(Mesomeric stabilisation)

7.3.2.2 Ortho-/para- directing deactivators

Halogen groups are unique in that they direct *ortho-/para-* and yet they deactivate
the benzene ring. The +M effect of Cl, Br and I is weak because these atoms are
all larger than carbon. This means that the orbitals containing the nonbonding
pairs of electrons (e.g. 3p-orbitals for Cl) do not overlap well with the 2p-orbital of
carbon. (This is a general phenomenon and mesomeric effects are not transmitted
well between atoms of different rows of the periodic table.) The weak +M effect
does, however, ensure that the halogens are *ortho-/para-* directing but the strong
–I effect (which deactivates the ring) is more significant in terms of the reactivity
of halobenzenes.

Notice the use of double-headed
resonance arrows (Section 1.6.3)

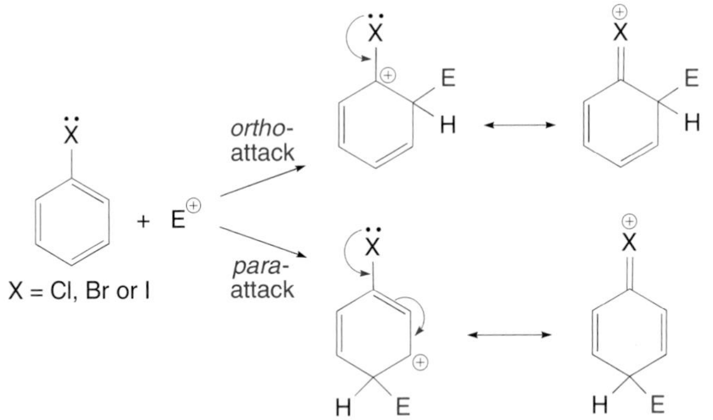

X = Cl, Br or I

ortho-
attack

para-
attack

7.3.2.3 Meta- directing deactivators

Electron withdrawing –I/–M groups (EWG), which make the ring less nucleophilic than benzene, will deactivate the *meta-* position less than the *ortho-/para-*
positions. The carbocation produced from attack at the *meta-* position will
therefore be the most stable because this does not reside adjacent to the EWG
in any of the resonance forms.

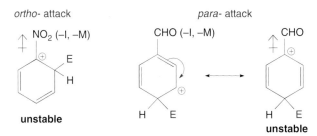

Examples

The EWG can destabilise both of these cations more effectively than the cation derived from *meta-* attack

NO$_2$ is a nitro group, while CHO is an aldehyde group (Section 2.1)

7.3.2.4   *Steric versus electronic effects*

With activating groups, we would expect the ratio of attack at the *ortho-* and *para-* positions to be 2:1 (as there are two *ortho-* positions to one *para-* position on the ring). However, attack at the *ortho-* position is often less than this because of *steric hindrance*. The size of the group on the benzene ring strongly influences the substitution at the adjacent *ortho-* position. In general, the larger the size of the group on the ring, the greater the proportion of *para-* substitution.

Steric hindrance is introduced in Section 4.4

For examples of steric hindrance in the synthesis of substituted benzenes, see Section 7.8

7.4   Nucleophilic aromatic substitution (the S$_N$Ar mechanism)

The reaction of a nucleophile with a benzene ring is very rare. This can only occur when strongly electron-withdrawing substituents (e.g. NO$_2$) are at the *ortho-* and/

General types of reaction are
introduced in Section 4.6

S_N2 reactions are discussed in
Section 5.3.1.1

or *para*-positions of aryl halides (ArX). The reaction proceeds by an addition-elimination mechanism and the electron-withdrawing groups (EWG) can stabilise the intermediate carbanion (called a Meisenheimer complex) by resonance. It should be noted that this is *not* an S_N2 reaction, as S_N2 reactions cannot take place at an sp^2 carbon atom (because the nucleophile cannot approach the C$-$X bond at an angle of 180°).

EWG = electron withdrawing group

For stabilisation of carbanions by
resonance, see Section 1.6.3

The order of reactivity of *aryl* halides is Ar$-$F $>>$ Ar$-$Cl $\sim$ Ar$-$Br $\sim$ Ar$-$I, which is the reverse order for S_N2/S_N1 reactions of *alkyl* halides (see Section 5.3.1). As fluorine is the most electronegative halogen atom, it strongly polarises the C$-$F bond. This makes the carbon atom δ+ and so it is susceptible to nucleophilic attack. The fluorine atom is also small and so the incoming nucleophile can easily approach the adjacent carbon atom (as there is little steric hindrance).

Steric hindrance is introduced in
Section 4.4

Aryl diazonium compounds (Ar$-$N$_2{}^+$) can also undergo substitution reactions in the presence of nucleophiles. However, the reactions appear to involve radical or ionic (S_N1-like) mechanisms. An S_N1 mechanism requires the formation of an aryl cation (Ar$^+$) which is highly unstable because the positive charge cannot be delocalised around the ring. It appears, however, that these highly reactive cations can be formed (and trapped by various nucleophiles) because gaseous nitrogen (N$_2$) is an excellent leaving group (Section 7.6).

S_N1 reactions are discussed in
Section 5.3.1.2

7.5 The formation of benzyne

In a synperiplanar elimination the
leaving groups lie in the same plane
on the same side of the molecule
(Section 6.2.1).

Reaction of halobenzenes (C$_6$H$_5$X) with a strong base (e.g. NaNH$_2$ in NH$_3$) leads to the elimination of HX and the formation of benzyne (in a synperiplanar elimination reaction). The new π-bond that is formed (by overlap of two sp^2

orbitals outside the ring) is very weak and benzyne can react with nucleophiles or take part (as the dienophile) in Diels-Alder reactions (see Section 6.2.2.11). As benzyne is extremely unstable it undergoes nucleophilic attack by $^-NH_2$ to give a new anion (this is not typical of ordinary alkynes).

The amide ion, $^-NH_2$, is a strong base (Section 1.7.6)

halobenzene

benzyne

aniline

For stabilisation of carbanions by resonance, see Section 1.6.3

7.6 Transformation of side chains

Functional groups on a benzene ring can generally be converted into other functional groups without any reaction of the stable aromatic ring.

1. *Oxidation of a methyl substituent.* Strong oxidising agents (such as potassium permanganate, $KMnO_4$) can convert a methyl group into a carboxylic acid (e.g. $PhCO_2H$).

toluene

benzoic acid

The CH_3 carbon atom in toluene is at oxidation level 0; the CO_2H carbon atom in benzoic acid is at oxidation level 3 (Section 4.5)

2. *Bromination of an alkyl substituent.* *N*-Bromosuccinimide (NBS) can brominate at the benzylic position (i.e. at the carbon atom attached to the benzene ring) via a radical chain mechanism. It should be noted that the intermediate benzylic radical is stabilised by resonance (i.e the radical can interact with the π electrons of the benzene ring).

NBS produces bromine at a controlled rate (Section 6.2.2.2)

Radical reactions are introduced in Section 4.6.2

For a related radical reaction involving a peroxide, see Section 6.2.2.1

benzylic radical

Catalytic hydrogenation of alkenes and alkynes is discussed in Section 6.3.2.4

3. *Reduction of nitro groups.* Catalytic hydrogenation or reaction with iron or tin in acid reduces an NO_2 group to an NH_2 group.

The deactivating NO_2 group is converted into an activating NH_2 group

4. *Reduction of ketone groups.* Catalytic hydrogenation or reaction with zinc amalgam in strong acid (in the *Clemmensen reduction*) reduces a –CO– group bonded to a benzene ring, to give a –CH_2– group.

The –CO– carbon atom in the ketone is at oxidation level 2; the –CH_2– carbon atom in the product is at oxidation level 0 (Section 4.5)

5. *The preparation and reactions of diazonium salts.* Reaction of aromatic amines ($ArNH_2$) with the nitrosonium ion (^+NO), generated from nitrous acid (HNO_2), yields aryl diazonium salts (ArN_2^+), which can be converted (on loss of nitrogen, N_2) into a range of functional groups.

Water is a good leaving group (Section 5.3.1.4)

When an H^+ ion moves to a different position within a molecule, this is called a proton transfer

This process is called *diazotisation*

On heating, the aryl diazonium ion (ArN_2^+) can produce nitrogen gas (an excellent leaving group) and a very unstable aryl cation (Ar^+) in an S_N1 mechanism, which can react with nucleophiles.

Nitrogen (N_2) is an excellent leaving group – it contains a strong triple bond and the loss of N_2 gas is favoured entropically

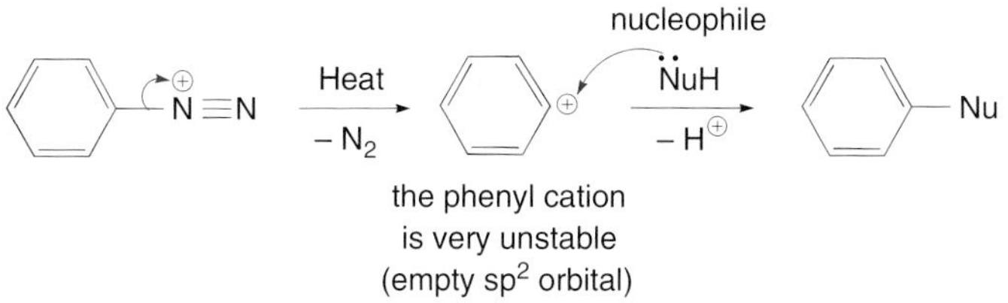

the phenyl cation
is very unstable
(empty sp^2 orbital)

An aryl cation (Ar^+) is very unstable because the vacant sp^2 orbital cannot overlap with the six p-orbitals of the aromatic ring

Functional group transformations

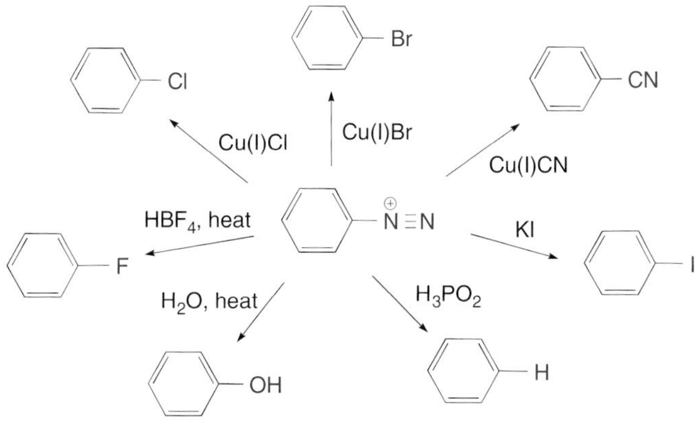

PhCN is called benzonitrile

H_3PO_2 is hypophosphorous acid; HBF_4 is fluoroboric acid

The reactions involving copper(I) salts are called *Sandmeyer reactions*. These reactions proceed via radical, rather than ionic, mechanisms.

Aryl diazonium salts (ArN_2^+) also undergo *coupling reactions* with phenol (PhOH) or aromatic amines (e.g. $ArNH_2$), which possess nucleophilic OH or NH_2 groups, respectively. This electrophilic substitution reaction (with the diazonium salt as the electrophile) produces highly coloured azo compounds.

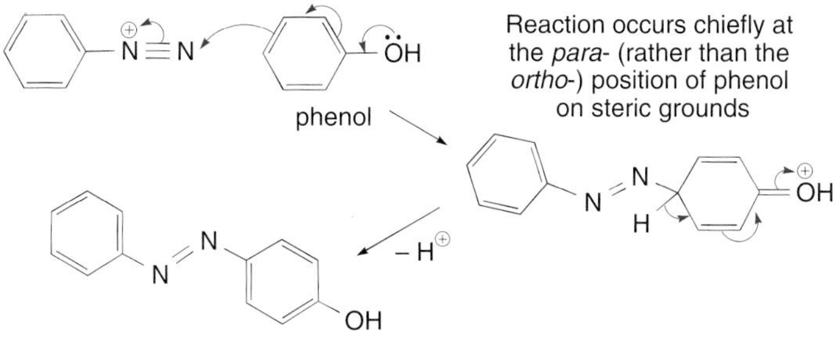

phenol

Reaction occurs chiefly at
the *para-* (rather than the
ortho-) position of phenol
on steric grounds

Steric effects and electrophilic substitution is discussed in Section 7.3.2.4

4-hydroxyazobenzene (orange)

7.7 Reduction of the benzene ring

Catalytic hydrogenation of alkenes and alkynes is discussed in Section 6.3.2.4

Harsh reaction conditions are required to reduce the aromatic benzene ring. This can be achieved by catalytic hydrogenation (using high temperatures or pressures and very active catalysts) or alkali metals in liquid ammonia/ethanol (in a *Birch reduction*).

Catalytic hydrogenation

In the chair conformation of a substituted cyclohexane, the substituent (R) sits in an equatorial position (Section 3.2.4)

A cyclohexane

Birch reduction

Addition of an electron to a neutral organic compound forms a radical anion; loss of an electron from a neutral organic compound forms a radical cation (Section 10.1.1)

Sodium or lithium metal (in liquid ammonia) can donate an electron to the benzene ring to form a radical anion ($R^{\bullet -}$). On protonation (by ethanol, EtOH) and further reduction then protonation, this produces 1,4-cyclohexadiene.

For a related reduction of an alkyne, see Section 6.3.2.4

radical anion   *radical*

carbanion

1,4-cyclohexadiene

7.8 The synthesis of substituted benzenes

The following points should be borne in mind when planning an efficient synthesis of a substituted benzene.

Activating groups donate electron density toward the benzene ring (Section 7.3.1)

1. The introduction of an *activating group* on to a benzene ring makes the product more reactive than the starting material. It is therefore difficult to stop after the first substitution. This is observed for Friedel-Crafts alkylations.

An alkyl group is an *ortho-/para-* directing activator

The mechanism of Friedel-Crafts alkylation is discussed in Section 7.2.4

An alkylbenzene is more reactive than benzene

Difficult to stop further alkylation (RCH_2Cl, $AlCl_3$)

+
ortho- isomer

2. The introduction of a *deactivating group* on to a benzene ring makes the product less reactive than the starting material. Therefore multiple substitutions do not occur, for example, in Friedel-Crafts acylations (to give ketones, e.g. ArCOR). The ketone (ArCOR) could then be reduced to give the mono-alkylated product (ArCH$_2$R) in generally higher yield than that derived from a Friedel-Crafts alkylation.

The mechanism of Friedel-Crafts acylation is discussed in Section 7.2.5

The Clemmensen reduction is introduced in Section 7.6

An acylbenzene is less reactive than benzene

3. For an *ortho-/para-* substituted benzene, the first group to be introduced on the ring should be *ortho-/para-* directing. For a *meta-* substituted benzene ring, the first group to be introduced on the ring should be *meta-* directing.

For the mechanism of nitration of a benzene ring, see Section 7.2.2

For the mechanism of bromination of a benzene ring, see Section 7.2.1

(*meta-* directing)

If the bromine atom had been introduced first then a mixture of 2- and 4-bromo-nitrobenzene would have been formed

3-bromonitrobenzene

4. For aniline (PhNH$_2$), the orientation of substitution depends on the pH of the reaction. At low pH aniline is protonated and the protonated amine is *meta-*directing.

The –NH$_3{}^+$ group has a –I effect and so it is a *meta*-directing deactivator (Section 7.3.2.3)

(*ortho-/para-* directing)

neutral pH

low pH

aniline

(*meta-* directing)

then base

Basicity of aniline and substituted anilines is discussed in Section 1.7.2

5. Making the activating group larger in size increases the proportion of *para-* over *ortho-* substitution (as steric hindrance minimises *ortho-* attack). For example, conversion of an amine (–NH$_2$) into a bulkier amide (–NHCOR) leads to selective *para*-substitution and the amide is described as a 'blocking

For the influence of steric effects on the position of electrophilic substitution in a substituted benzene ring see Section 7.3.2.4

group'. The amide group is also not as strong an activating group as the amine (see Section 7.3.1), and hence multiple substitution is less of a problem.

Formation of amides from acyl chlorides is introduced in Section 9.5, while amide hydrolysis is discussed in Section 9.8

Reaction of aniline with an electrophile gives a mixture of *ortho-* and *para-* isomers

6. Removable aromatic substituents, typically SO_3H or NH_2, can be used to direct or block a particular substitution. In this way, aromatic products with unusual substitution patterns can be prepared.

The $-SO_3H$ group can be removed using hot, dilute aqueous acid (Section 7.2.3)

Reaction of the substituted aniline with $NaNO_2$ and HCl forms an intermediate diazonium ion (Section 7.6)

Because the NH_2 group in aniline is strongly activating, for bromination, a Lewis acid catalyst is not required

The amine group is removed leaving three Br groups *meta-* to one another

7. Deactivating groups can be converted into activating groups (and vice versa) by functional group interconversion.

For directing group effects see Section 7.3.2

8. For the formation of trisubstituted benzenes, the directing effects of the two groups on the benzene ring must be compared. When these groups direct to different positions, the more powerful activating group usually has the dominant influence. Substitution in between two *meta-*disubstituted groups is rare because of steric hindrance.

Steric hindrance is discussed in Section 7.3.2.4, and introduced in Section 4.4

(*ortho-/para-* directing)

CH₃

E⊕

CO₂H

(*meta-* directing)

(*ortho-/para-* directing)

OH

E⊕

CH₃

(*ortho-/para-* directing)

OH is a more powerful activating group

Attack at this site is hindered

(*ortho-/para-* directing)

CH₃

E⊕

Cl

(*ortho-/para-* directing)

E⊕

4-Methylbenzoic acid is otherwise known as *p*-toluic acid

4-Methylphenol is otherwise known as *p*-cresol

7.9 Electrophilic substitution of naphthalene

As naphthalene ($C_{10}H_8$) is aromatic (10 π electrons), it also undergoes electrophilic substitution reactions. Attack occurs selectively at the C-1 position, rather than the C-2 position, because the intermediate carbocation is more stable (i.e. two resonance structures can be drawn with an intact benzene ring; for attack at C-2, only one can be drawn with an intact benzene ring).

Naphthalene is an example of a polycyclic aromatic hydrocarbon (PAH)

For sulfonation (using H_2SO_4), the position of attack depends on the reaction temperature. At 80 °C, attack at C-1 occurs (as expected), and the 1-sulfonic acid is formed under kinetic control (Section 4.9.3). However, at higher temperatures (e.g. 160 °C), attack at C-2 predominates and the 2-sulfonic acid is formed under thermodynamic control. The C-1 isomer is thermodynamically less stable than the C-2 isomer because of an unfavourable steric interaction with the hydrogen atom at C-8 (this is called a *peri interaction*).

For the mechanism of sulfonation of benzene see Section 7.2.3

7.10 Electrophilic substitution of pyridine

Pyridine (C_5H_5N), like benzene, has 6 π electrons. The electron withdrawing nitrogen atom deactivates the ring and electrophilic substitution is slower than that for benzene. Substitution occurs principally at the 3-position of the ring, as attack at the 2-/4-positions produce less stable cation intermediates (i.e. with one resonance structure having a positive charge on the divalent nitrogen).

Pyridine is commonly used as a base in organic synthesis (Section 1.7.5). The lone pair of electrons on nitrogen is not part of the aromatic ring

Attack at C-2 Unfavourable

Attack at C-4 Unfavourable

Attack at C-3 **Preferred**

Nitrogen lone pair is not part of the aromatic sextet

7.11 Electrophilic substitution of pyrrole, furan and thiophene

In pyrrole (C_4H_4NH), the lone pair of electrons on nitrogen is part of the aromatic (six π-electron) ring system. The incorporation of a lone pair of electrons activates the ring, and electrophilic substitution is faster than that for benzene. Substitution occurs principally at the 2-position, as attack at the 3-position produces a less stable cation intermediate (i.e. with two, rather than three, resonance structures).

For stabilisation of carbocations by resonance, see Section 1.6.3

As the lone pair is part of the aromatic sextet it is less nucleophilic/basic than for aliphatic amines

C-2 substituted pyrrole

Pyrrole is much less basic than pyridine (Section 1.7.5).

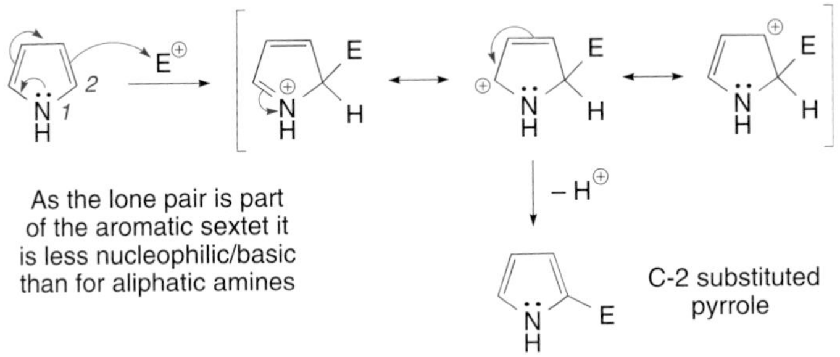

Furan (C_4H_4O) and thiophene (C_4H_4S) also undergo electrophilic substitution reactions although not so readily as pyrrole. The typical order of reactivity to electrophiles is: pyrrole > furan > thiophene > benzene. Pyrrole is most reactive because the nitrogen atom is a more powerful electron donor than the oxygen atom in furan (i.e. nitrogen is less electronegative than oxygen). Thiophene is less reactive than either pyrrole or furan because the lone pair of electrons on sulfur are in a 3p-orbital (rather than a 2p-orbital). The 3p-orbitals of sulfur overlap less efficiently with the 2p-orbitals on carbon than the (similar size) 2p-orbitals of nitrogen or oxygen. As for pyrrole, the 2-position of both furan and thiophene is more reactive to electrophiles than the 3-position.

Pyrrole, furan, thiophene and pyridine are all examples of heteroaromatic compounds

Examples

For the mechanism of bromination of benzene, see Section 7.2.1

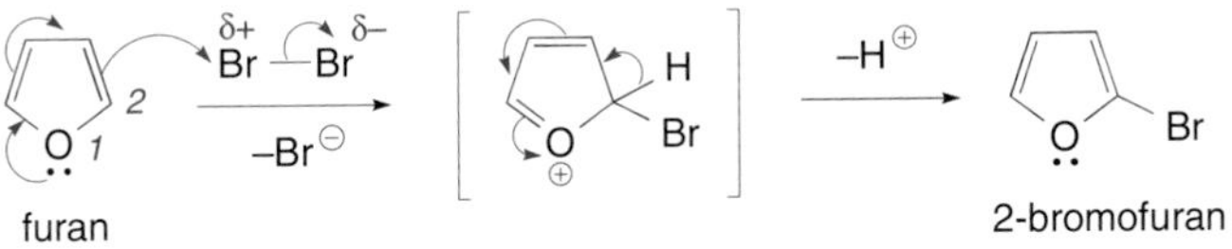

furan

2-bromofuran

Furan is more reactive to electrophiles than benzene and so $FeBr_3$ is not required

For the mechanism of Friedel-Crafts acylation of benzene, see Section 7.2.5

The IUPAC name for COMe is ethanoyl

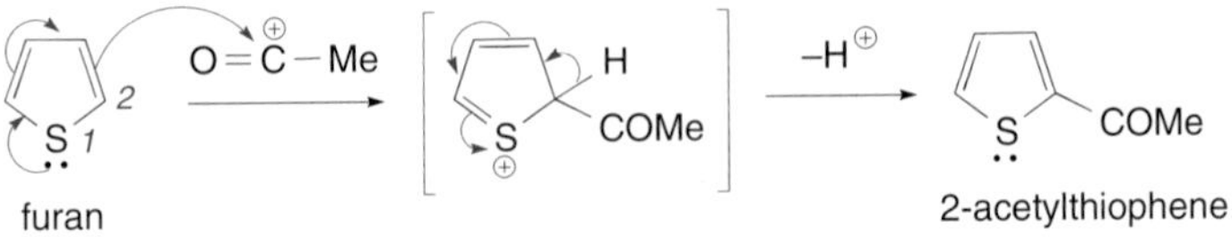

furan

2-acetylthiophene

The acylium ion, $MeC^+=O$, can be prepared using MeCOCl and $SnCl_4$

Worked example

(a) Giving your reasons, draw the major product from each of the following reactions.

(i)

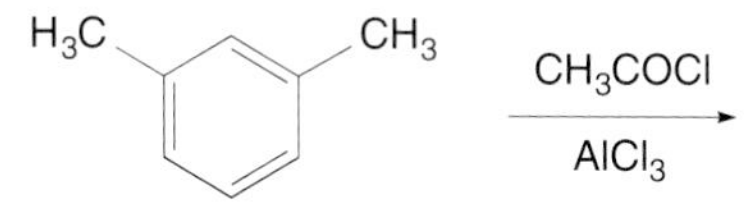

CH_3COCl
$\xrightarrow{\quad\quad}$
$AlCl_3$

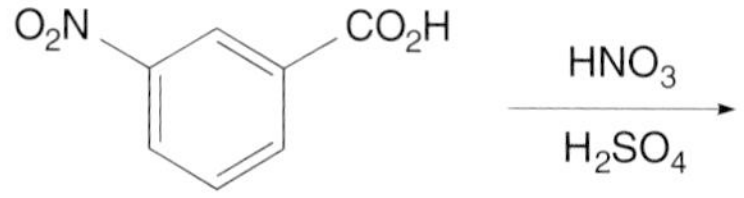

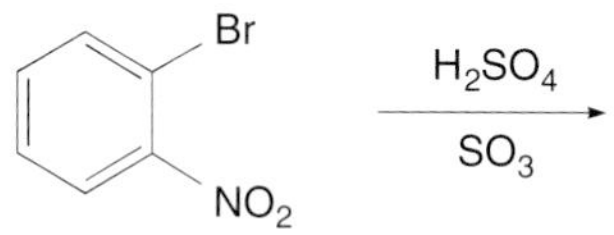

Hint: Consider the electrophiles that are formed from the reagents. For all substituents on the ring (of the precursors), assign directing effects, and consider steric effects

(ii)

HNO_3
$\xrightarrow{\quad\quad}$
H_2SO_4

(iii)

H_2SO_4
$\xrightarrow{\quad\quad}$
SO_3

(b) Suggest an efficient synthesis of the compound shown below, starting from phenol.

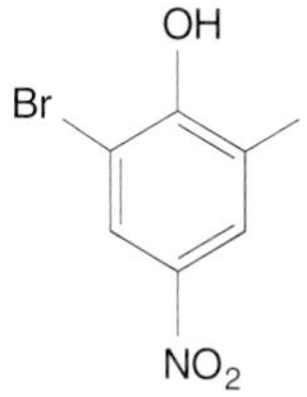

Hint: For the substituents on the ring, assign directing effects to determine the order of electrophilic substitution

Answer

(a)

(i)

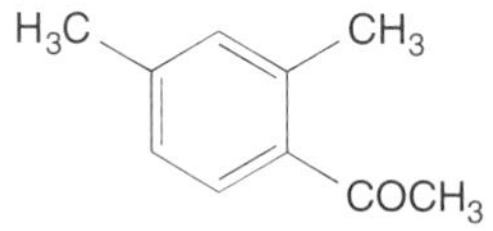

The methyl groups are 2,4-directing and the electrophile, $CH_3C^+=O$, is introduced at the least hindered position

For Friedel-Crafts acylation of benzene, see Section 7.2.5

(ii)

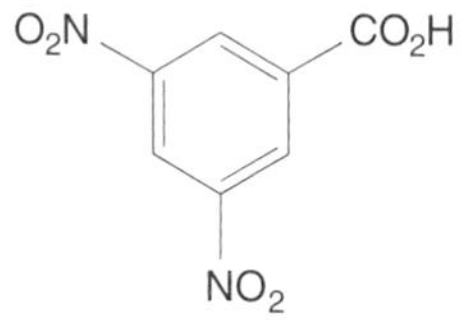

The NO_2 and CO_2H groups are both 3-directing; the electrophile is $^+NO_2$

For nitration of benzene see Section 7.2.2

(iii)

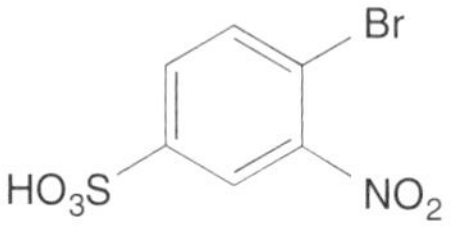

The Br group is 2,4-directing and the NO_2 group is 3-directing; the electrophile is $^+SO_3H$

For bromination of benzene see Section 7.2.1

(b)

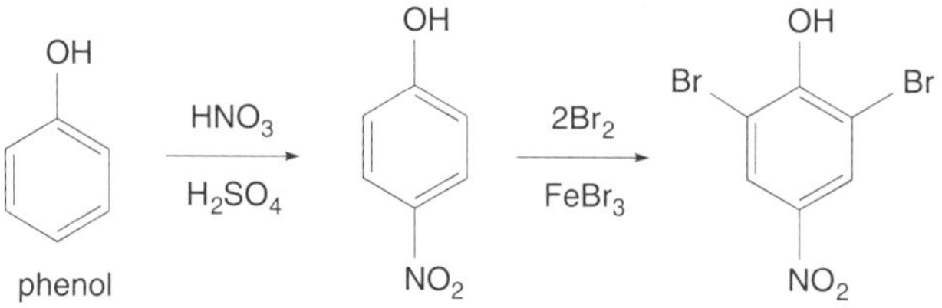

HNO_3
$\xrightarrow{\quad\quad}$
H_2SO_4

$2Br_2$
$\xrightarrow{\quad\quad}$
$FeBr_3$

phenol

+ 2-nitrophenol (separate)

Phenol is introduced in Section 2.4

Directing group effects are discussed in Section 7.3.2

Problems

1. Give the mechanism of the electrophilic substitution reaction of benzene with an electrophile E^+.

Section 7.2

Section 7.2.4

2. Draw a mechanism for the formation of (1,1-dimethylpropyl)benzene from 1-chloro-2,2-dimethylpropane, benzene and a catalytic amount of AlCl$_3$.

Section 7.1

3. (a) On the basis of Hückel's rule, label the following molecules **A**–**D** as aromatic or anti-aromatic.

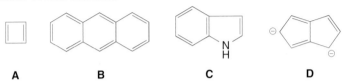

    A   **B**   **C**   **D**

Section 7.2

  (b) Describe one experimental approach for testing the aromatic character of **C**.

Sections 7.2, 7.3 and 7.8

4. Suggest syntheses of the following compounds starting from benzene.

  (a) Triphenylmethane

  (b) (1-Bromopropyl)benzene

  (c) 4-Bromobenzoic acid

  (d) 2,4-Dinitrophenylhydrazine

Sections 7.2, 7.3 and 7.8

5. Explain the following:

  (a) Reaction of phenylamine (aniline) with bromine yields 2,4,6-tribromophenylamine whereas nitration (with a mixture of concentrated nitric and hydrochloric acids) gives mainly 3-nitrophenylamine.

  (b) Bromination of benzene with Br$_2$ requires FeBr$_3$, whereas reaction of Br$_2$ with cyclohexene or phenol does not.

  (c) Although 4-bromophenylamine can be formed in one step from phenylamine, the yield is low and so a 3-step synthesis involving the formation of an intermediate amide is preferred.

Section 7.6

6. How would you prepare the following compounds from phenylamine (aniline) using arenediazonium salts as intermediates?

  (a) Iodobenzene

  (b) 4-Bromochlorobenzene

  (c) 1,3,5-Tribromobenzene

7. Draw mechanisms for the following reactions.

Section 7.5

(a)   OMe, Cl $\xrightarrow[\text{NH}_3]{\text{NaNH}_2}$ OMe, NH$_2$

Section 7.4

(b)   F, NO$_2$, NO$_2$ $\xrightarrow{\text{NaOH}}$ OH, NO$_2$, NO$_2$

8

Carbonyl compounds: aldehydes and ketones

Key point. The carbonyl (C=O) bond is polarised, and the oxygen atom is slightly negative, while the carbon atom is slightly positive. This explains the addition of nucleophiles to the carbon atom of aldehydes (RCHO) or ketones (RCOR) in *nucleophilic addition reactions*. A variety of charged or neutral nucleophiles can add to the carbonyl group, although addition of neutral nucleophiles usually requires the presence of an acid catalyst. Aldehydes and ketones bearing α-hydrogen atoms can undergo tautomerism to form *enols* or, on reaction with a base, undergo deprotonation to form *enolate ions (enolates)*. Enols, and particularly enolates, can act as nucleophiles and reaction with an electrophile leads to an *α-substitution reaction*. When an enolate reacts with an aldehyde or ketone this can produce an *enone* in a *carbonyl-carbonyl condensation reaction*.

nucleophilic addition

enol form

enolate ion

8.1 Structure

All carbonyl compounds contain an acyl fragment (RC=O) bonded to another residue. The carbonyl carbon atom is sp^2 hybridised (three σ-bonds and one π-bond) and, as a consequence, the carbonyl group is planar and has bond angles of around 120°. The C=O bond is short (1.22 Å) and also rather strong (ca. $690 \, kJ \, mol^{-1}$).

Hybridisation is discussed in Section 1.5

Keynotes in Organic Chemistry, Second Edition. Andrew F. Parsons.
© 2014 John Wiley & Sons, Ltd. Published 2014 by John Wiley & Sons, Ltd.

Naming of aldehydes and ketones
is discussed in Section 2.4

π bond

two lone
pairs

Ketone: R = R^1 = alkyl or aryl
Aldehyde: R = alkyl or aryl, R^1 = H

PhCOCH$_3$ is an aromatic ketone,
CH$_3$COCH$_3$ is an aliphatic ketone,
and cyclohexanone, (CH$_2$)$_5$CO, is
an alicyclic ketone (Section 5.1)

As oxygen is more electronegative than carbon, the electrons in the C=O
bond are drawn towards the oxygen. This means that carbonyl compounds are
polar and have substantial dipole moments.

Carbonyl compounds show characteristic peaks in the infrared (IR) and ^{13}C
NMR spectra (see Sections 10.4 and 10.5.2). A characteristic signal is also
observed for the aldehyde hydrogen, RCHO, in the ^{1}H NMR spectrum (see
Section 10.5.1).

8.2 Reactivity

For the Pauling electronegativity
scale see Section 1.6.1

Polarisation of the C=O bond means that the carbon atom is electrophilic (δ+)
and the oxygen atom is nucleophilic (δ−). Therefore nucleophiles attack the
carbon atom and electrophiles attack the oxygen atom.

Hybridisation is discussed in
Section 1.5

sp^2 carbon   sp^3 carbon

either face of the C=O bond is
equally likely to be attacked

σ- and π-bonds are discussed in
Section 1.4

Attack by a nucleophile breaks the π-bond, and the electrons reside on the
oxygen atom. This is energetically favourable, as a strong σ-bond is formed at the
expense of a weaker π-bond. Examination of crystal structures has shown that the
nucleophile approaches the carbonyl carbon at an angle of around 107°. This is
called the *Bürgi-Dunitz angle* (see Section 4.10).

Nucleophilic addition reactions are
introduced in 4.6.1.1

There are three general mechanisms by which aldehydes and ketones react.
These are (i) *nucleophilic addition reactions*, (ii) *α-substitution reactions* and (iii)
carbonyl-carbonyl condensation reactions.

Nucleophiles and nucleophilic
strength is discussed in Section
4.2.1.1

1. *Nucleophilic addition reactions.* Both charged and uncharged nucleophiles can
   attack the carbonyl carbon atom to form addition products. This is the most
   common reaction for aldehydes and ketones.

With charged nucleophiles

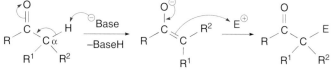

Nucleophiles = $H^{\ominus}$, $R^{\ominus}$, $^{\ominus}CN$

With uncharged nucleophiles

Nucleophiles = H_2O, ROH, RSH, NH_3, RNH_2, R_2NH
This process can be acid catalysed

Alcohols (ROH), thiols (RSH) and amines (e.g. RNH_2) are introduced in Section 2.1

Proton transfers involve movement of H^+ from one part of the molecule to another

1. *α-Substitution reactions.* This involves reaction at the position next to the carbonyl group, which is called the α–position. Deprotonation produces an enolate ion, which can act as a nucleophile.

The carbonyl group renders the hydrogen(s) on the α-carbon acidic

An enolate ion

Formation of an enolate ion is an example of an acid-base reaction (Section 1.7.6)

2. *Carbonyl-carbonyl condensation reactions.* These reactions involve both a nucleophilic addition step and an α–substitution step. They can occur when two carbonyl compounds react with one another. For example, reaction of two molecules of ethanal (CH_3CHO) forms a β-hydroxyaldehyde (or aldol):

The addition product contains both an aldehyde and an alcohol, hence the name aldol

If two molecules of a ketone react, a β-hydroxyketone is formed; the name aldol reaction is also used to describe this process

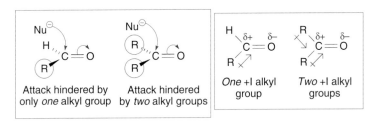

ethanal

nucleophile (enolate ion)

electrophile

In a condensation reaction, two molecules combine to give a product with loss of a small molecule, such as water

β-hydroxyaldehyde (aldol)

new C–C bond

The aldol addition product can subsequently undergo elimination of water to form an enal (see Section 8.5.1).

8.3 Nucleophilic addition reactions

8.3.1 Relative reactivity of aldehydes and ketones

Aldehydes (RCHO) are generally more reactive than ketones (RCOR) for both steric and electronic reasons.

1. *On steric grounds.* The nucleophile can attack the aldehyde carbonyl carbon atom more readily as this has only one (rather than two) alkyl group(s) bonded to it. The transition-state resulting from addition to the aldehyde is less crowded and lower in energy.

Steric effects are introduced in Section 4.4

Positive inductive (+I) effects are discussed in Section 1.6.1

Attack hindered by only *one* alkyl group

Attack hindered by *two* alkyl groups

One +I alkyl group

Two +I alkyl groups

2. *On electronic grounds.* Aldehydes have a more electrophilic carbonyl carbon atom because there is only one (rather than two) alkyl group(s) that donates electron density towards it.

Electrophiles and electrophilic strength is discussed in Section 4.2.1.2

8.3.2 Types of nucleophiles

Nucleophiles and nucleophilic strength is discussed in Section 4.2.1.1

The attacking nucleophile can be negatively charged (Nu^-) or neutral (Nu). With neutral nucleophiles, acid catalysis is common.

Nu$^-$ = H$^-$ (hydride), R$^-$ (carbanion), NC$^-$ (cyanide), HO$^-$ (hydroxide), RO$^-$ (alkoxide).

Nu = H$_2$O (water), ROH (alcohol), RSH (thiol), NH$_3$ (ammonia), RNH$_2$ (primary amine) or R$_2$NH (secondary amine).

8.3.3 Nucleophilic addition of hydride: reduction

Reduction of aldehydes (RCHO) and ketones (RCOR) with hydride ions (H$^-$) leads to alcohols. Primary alcohols (RCH$_2$OH) are formed from reduction of aldehydes, and secondary alcohols (R$_2$CHOH) from reduction of ketones. The hydride ion can be derived from a number of reagents.

Oxidation levels can be used to assess if a reaction involves reduction (Section 4.5)

8.3.3.1 Complex metal hydrides

Lithium aluminium hydride (LiAlH$_4$) or sodium borohydride (NaBH$_4$) can act as hydride donors. A *simplified* view of the mechanism involves the formation of an intermediate tetrahedral alkoxide ion, which on protonation yields a primary or secondary alcohol.

The protonation step is often described as a 'work-up' – a reaction carried out after the main reaction, leading to isolation and purification of the product

primary alcohol

secondary alcohol

Either H$^+$ or H$_3$O$^+$ could be shown as the proton donor

The metal–hydrogen bonds in LiAlH$_4$ are more polar than those in NaBH$_4$ (Al is more electropositive than B). As a consequence, LiAlH$_4$ is a stronger reducing agent than NaBH$_4$.

8.3.3.2 The Meerwein-Ponndorf-Verley reaction

The hydride ion is derived from 2-propanol (Me$_2$CHOH), which is first deprotonated to form an alkoxide ion.

The pK_a of an alcohol is 16–17 (Appendix 3)

Equilibria is discussed in Section 4.9.1.1

Propanone (acetone) is widely used as a solvent

The equilibrium is shifted to the (secondary) alcohol if propanone (b.p. 56 °C) is removed by distillation

Lewis acids are introduced in Section 1.7.3

Other methods for oxidising alcohols, to form aldehydes or ketones, are discussed in Section 8.3.3.5

Lewis acids, such as Al(OR)$_3$, are employed so as to facilitate hydride transfer by forming a complex with the alkoxide ion and carbonyl compound.

The reverse reaction, involving the oxidation of the secondary alcohol to form a ketone, is known as the *Oppenauer oxidation*. In this case, the equilibrium is pushed to the ketone by using excess propanone.

8.3.3.3 The Cannizzaro reaction

The hydride ion is derived from aldehydes such as methanal (HCHO), which do not contain an α-hydrogen atom (i.e. a hydrogen atom on the carbon atom next to the carbonyl).

Carboxylic acids (RCO$_2$H) are stronger acids than alcohols (ROH). Deprotonation of a carboxylic acid forms a carboxylate ion, RCO$_2^-$, that is stabilised by resonance (Section 1.7.1)

This is a *disproportionation reaction*, as one molecule of methanal is oxidised to methanoate (which on protonation gives methanoic acid, HCO_2H) and the other molecule is reduced to methanol.

8.3.3.4 Nicotinamide adenine dinucleotide (NADH)

Similar hydride reductions occur in nature using NADH in the presence of enzymes.

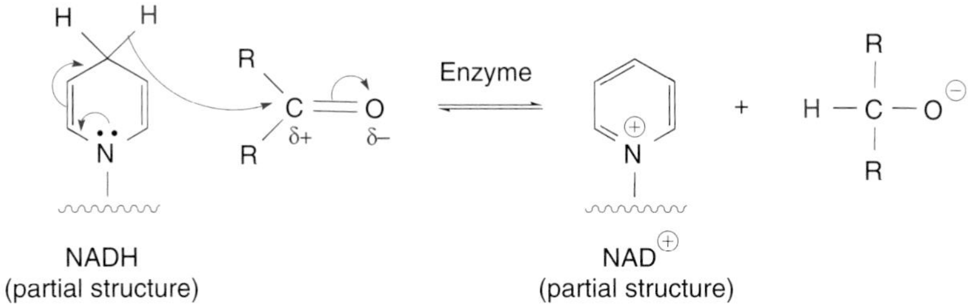
Enzyme
NADH
(partial structure)
NAD⊕
(partial structure)

8.3.3.5 The reverse of hydride addition: oxidation of alcohols

The conversion of *secondary* alcohols (R_2CHOH) into ketones (RCOR) is achieved using $KMnO_4$, $Na_2Cr_2O_7$ or CrO_3 under acidic conditions. The use of CrO_3 in acid is called the *Jones oxidation*.

Oxidation levels can be used to assess if a reaction involves oxidation (Section 4.5)

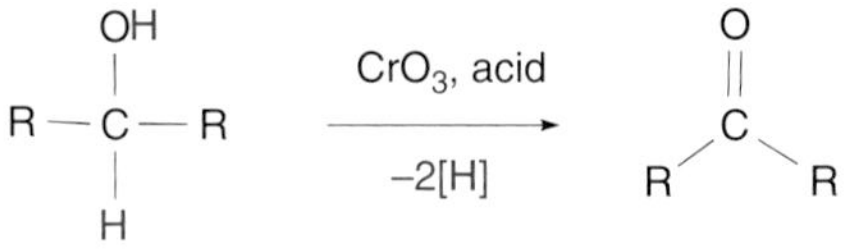
CrO3, acid
–2[H]

The mechanism involves the formation of a chromate ester and, at the end of the reaction, the oxidation state of the chromium changes from +6 (orange colour) to +3 (green colour), i.e. the chromium is reduced.

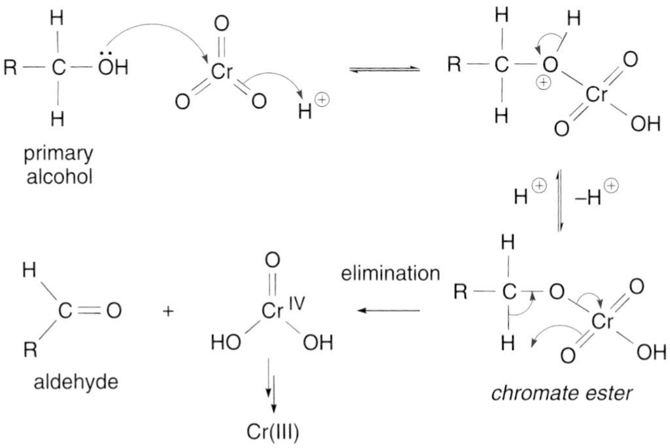
primary
alcohol
elimination
aldehyde
Cr(III)
chromate ester

Related Ei eliminations (to form alkenes) are discussed in Section 6.2.1

For *primary* alcohols (RCH_2OH), it is often difficult to stop at the aldehyde stage and further oxidation to give the carboxylic acid (RCO_2H) can occur. The aldehyde can be isolated by distillation (i.e. removed from the reaction mixture as soon as it is formed).

Reactions of carboxylic acids are discussed in Section 9.4

primary alcohol → CrO₃, acid oxidation → aldehyde can be isolated by distillation → CrO₃, acid oxidation → carboxylic acid

Oxidising agents milder than CrO_3/H^+, including pyridinium chlorochromate ($C_5H_5NH^+$ $ClCrO_3{}^-$), abbreviated as PCC, can also be used. PCC is an excellent reagent for oxidising primary alcohols into aldehydes in anhydrous dichloromethane (CH_2Cl_2).

Alcohols can be formed by addition of water to alkenes (Sections 6.2.2.4 and 6.2.2.5)

Tertiary alcohols ($R_3C–OH$) cannot be oxidised by CrO_3/H^+ (or related reagents) because there are no hydrogen atoms on the carbon atom bearing the OH group. As a result, oxidation must take place by breaking carbon–carbon bonds (e.g. in combustion, to give CO_2 and H_2O).

8.3.4 Nucleophilic addition of carbon nucleophiles: formation of C–C bonds

8.3.4.1 Reaction with cyanide

Catalysts are discussed in Section 4.9.2

Addition of cyanide (^-CN) leads to the reversible formation of a cyanohydrin (e.g. $R_2C(OH)CN$). Only a catalytic amount of the cyanide ion in the presence of hydrogen cyanide is required (as the cyanide ion is regenerated).

For ketones, RCOR, with large alkyl groups, the equilibrium favours the ketone rather than the cyanohydrin

The position of the equilibrium depends on the structure of the carbonyl compound. Nucleophilic addition is favoured by small alkyl groups (R) attached to the carbonyl group, and also by electron-withdrawing groups (e.g. CCl_3), which increase the δ+ character of the carbonyl carbon atom (Section 8.3.1).

Cyanohydrins are useful because they can be converted into other functional groups, e.g. hydroxy acids (see Section 9.9 for the reaction mechanism).

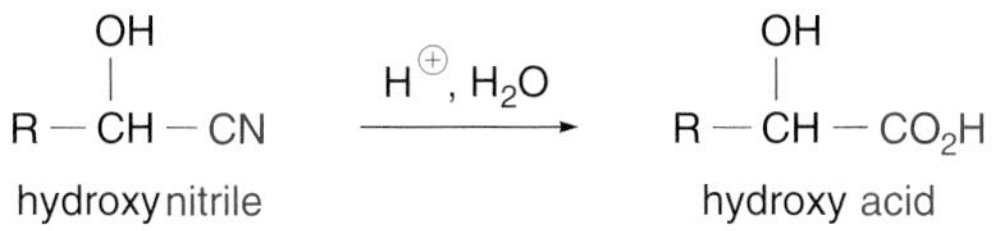
OH
R — CH — CN
hydroxynitrile
H⊕, H2O
OH
R — CH — CO2H
hydroxy acid

The reaction of cyanide with *aromatic* aldehydes leads to the *benzoin condensation reaction*. The product from this reaction is called benzoin.

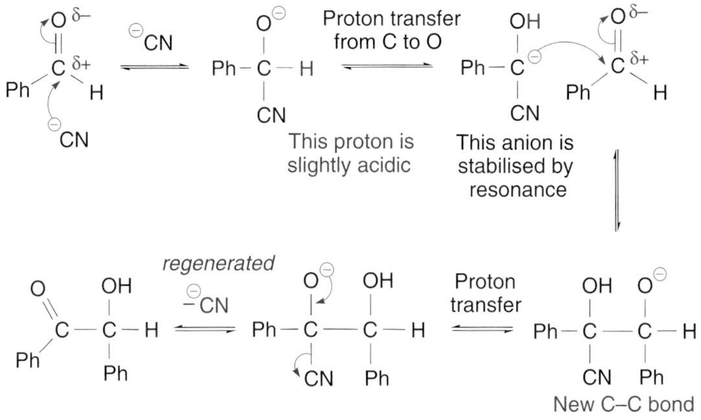
Proton transfer
from C to O
This proton is
slightly acidic
This anion is
stabilised by
resonance
regenerated
Proton
transfer
New C–C bond

PhCHO is called benzaldehyde

For resonance stabilisation of anions see Section 1.6.3

8.3.4.2 Reaction with organometallics

Organometallic compounds (R–Metal) are a source of nucleophilic alkyl or aryl groups. This is because the metal is more electropositive than the carbon atom to which it is bonded.

Organometallics contain an organic group bonded to a metal

$$\overset{\delta-}{\mathrm{R}} \text{———} \overset{\delta+}{\mathrm{Metal}}$$

Organometallic reagents include:

1. organolithium reagents, R–Li;
2. Grignard reagents, R–MgX (where X = Cl, Br or I);
3. the Reformatskii reagent, BrZn–CH_2CO_2Et;
4. alkynylmetal reagents, e.g. RC≡C^- Na^+.

The nucleophilic alkyl or aryl group adds to the carbonyl to form an alcohol. These reactions are very useful in the synthesis of complex organic molecules because new carbon–carbon bonds are formed.

Notice that in this reaction, a ketone is converted into a tertiary alcohol

Examples

For a Grignard reaction, the byproduct is HOMgX

Organolithium and Grignard reagents are prepared from alkyl or aryl halides, RX or ArX. The reactions are carried out under anhydrous conditions because reaction with water leads to alkanes.

Reaction of a Grignard reagent with water is an example of an acid–base reaction (Section 1.7.6)

$$R-X \quad + \quad 2Li \xrightarrow{\text{dry pentane}} R-Li \quad + \quad LiX$$

$$R-X + Mg \xrightarrow{\substack{\text{dry diethyl} \\ \text{ether (EtOEt)}}} R-MgX \xrightarrow{H-OH} R-H + HOMgX$$
alkane

8.3.4.3   Reaction with phosphonium ylides: the Wittig reaction

Alkenes are also prepared using elimination reactions (Section 6.2.1)

Aldehydes (RCHO) and ketones (RCOR) react with phosphonium ylides or phosphoranes ($R_2C=PPh_3$) to form alkenes. The first step involves nucleophilic attack by the carbon atom of the phosphonium ylide.

A ylide contains an anionic site attached directly to a positively charged heteroatom

Notice that phosphorus can accommodate 10 electrons in its valence shell

The driving force for the reaction is the formation of the very strong P=O bond in triphenylphosphine oxide, Ph_3PO.

Phosphonium ylides are prepared from a nucleophilic substitution reaction between halogenoalkanes (RX) and triphenylphosphine (PPh_3). The resulting alkyltriphenylphosphonium salt is then deprotonated by reaction with a base to form the phosphonium ylide.

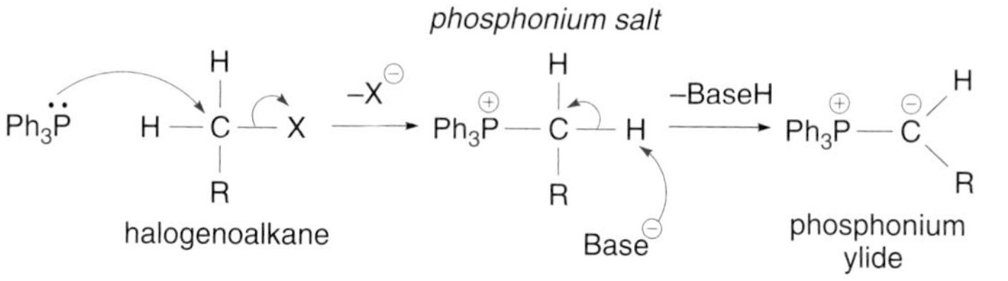

phosphonium salt

halogenoalkane

Base

phosphonium ylide

If RX is a primary halogenoalkane, then, as shown, the first step is an S_N2 reaction (Section 5.3.1.1)

8.3.5 Nucleophilic addition of oxygen nucleophiles: formation of hydrates and acetals

8.3.5.1 Addition of water: hydration

Aldehydes (RCHO) and ketones (RCOR) undergo a reversible reaction with water to yield hydrates, $RCH(OH)_2$ or $R_2C(OH)_2$ (otherwise known as geminal (gem) diols or 1,1-diols).

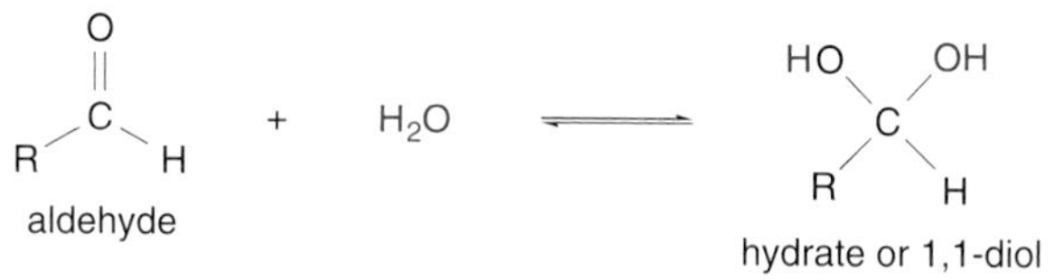

aldehyde

hydrate or 1,1-diol

The addition of water is slow but can be catalysed by bases or acids.

Catalysts are introduced in Section 4.9.2

Base catalysis

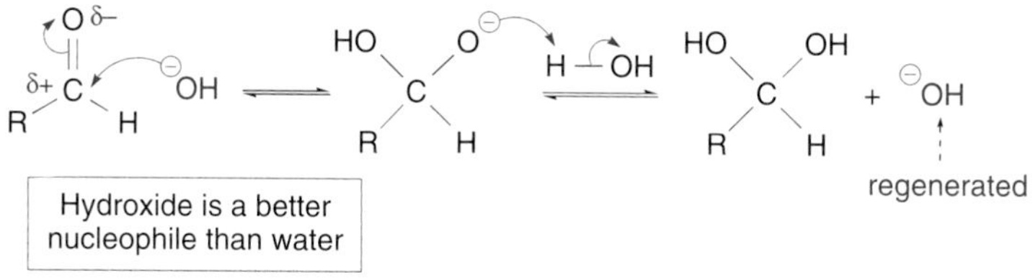

Hydroxide is a better nucleophile than water

regenerated

Nucleophilic strength is discussed in Section 4.2.1.1

Acid catalysis

Electrophilic strength is discussed
in Section 4.2.1.2

The carbonyl
acts as a base

The protonated carbonyl
is a better electrophile

regenerated

For simplicity the H$^+$ is shown
dropping off the oxygen, but in the
reaction mixture a base is needed to
remove the H$^+$ (such as H$_2$O)

Electron-donating and bulky substituents attached to the carbonyl group decrease the percentage of hydrate present at equilibrium, whereas electron-withdrawing and small substituents increase it. Therefore, only 0.2% of propanone (O=CMe$_2$) is hydrated at equilibrium, while 99.9% of methanal (O=CH$_2$) is hydrated.

8.3.5.2 Addition of alcohols: hemiacetal and acetal formation

Alcohols (ROH) are relatively weak nucleophiles (like water) but add rapidly to aldehydes (RCHO) and ketones (RCOR) on acid catalysis. The initial product is a hydroxyether or *hemiacetal*, RCH(OH)OR or R$_2$C(OH)OR.

Notice that, as for hydrate
formation (Section 8.3.5.1), the
formation of a hemiacetal is
reversible

aldehyde

hemiacetal

Hemiacetal formation is fundamental to the chemistry of carbohydrates (see Section 11.1). Glucose, for example, contains an aldehyde and several alcohol groups. Reaction of the aldehyde with one of the alcohols leads to the formation of a cyclic hemiacetal (even without acid catalysis) in an intramolecular reaction.

An intramolecular reaction involves
the reaction between two or more
sites within the same molecule

glucose

Proton
transfer

cyclic hemiacetal

Hemiacetals can undergo further reaction, with a second equivalent of alcohol (ROH), to yield an *acetal*, RCH(OR)$_2$ or R$_2$C(OR)$_2$.

Notice that protonation of the OH group converts it into a good leaving group, namely water (Section 5.3.1.4)

The whole sequence of acetal formation is *reversible*.

To prepare an acetal, a dehydrating agent (e.g. MgSO$_4$) is added; this removes the water from the reaction mixture as soon it is formed

Acetals are extremely useful compounds in synthesis because they can act as *protecting groups* for aldehydes and ketones. Acetals are less reactive than carbonyl compounds to nucleophiles and reducing agents, such as lithium aluminium hydride (LiAlH$_4$).

Example

Ethane-1,2-diol, HOCH$_2$CH$_2$OH, is an inexpensive and readily available diol

For the reduction of esters using LiAlH$_4$, see Section 9.7

8.3.6 Nucleophilic addition of sulfur nucleophiles: formation of thioacetals

Thiols (RSH) add reversibly to aldehydes (RCHO) and ketones (RCOR) in the presence of an acid catalyst to yield *thioacetals*, RCH(SR)$_2$ or R$_2$C(SR)$_2$. The reaction mechanism is analogous to the formation of acetals (see Section 8.3.5.2).

Thiols (RSH) are stronger nucleophiles than alcohols (ROH). Sulfur is larger than oxygen and its lone pairs of electrons are accessible, and most easily donated to an electrophile

Thioacetals are useful in synthesis because they undergo desulfurisation when treated with Raney nickel in the *Mozingo reduction*. This is an excellent two-step method for reducing aldehydes (RCHO) or ketones (RCOR) into alkanes (RCH$_3$ or RCH$_2$R).

Raney nickel is a finely powdered nickel-aluminium alloy that contains hydrogen absorbed onto the surface

Example

cyclohexanone   cyclic thioacetal   cyclohexane

8.3.7 Nucleophilic addition of amine nucleophiles: formation of imines and enamines

Amines are introduced in Section 2.1. Amines (e.g. RNH$_2$) are stronger nucleophiles than alcohols (ROH) as nitrogen is less electronegative than oxygen (Section 4.2.1.1)

Aldehydes (RCHO) and ketones (RCOR) react with primary amines (RNH$_2$) to yield *imines* (RCH = NR or R$_2$C = NR) and with secondary amines (R$_2$NH) to form *enamines* (e.g. RCH = CH–NR$_2$).

8.3.7.1 Formation of imines

This reversible, acid-catalysed process involves the addition of a primary amine nucleophile (RNH$_2$) followed by elimination of water; this is called an *addition-elimination* or *condensation* reaction. (A condensation reaction involves the combination of two molecules to form one larger molecule, together with the elimination of a small molecule, often water.)

Amines are strong nucleophiles that add to aldehydes and ketones in the absence of an acid catalyst, but an acid catalyst is required for the elimination step

The reaction is pH dependent.

- At *low* pH (strongly acidic conditions), the amine is protonated (RNH_3^+) and so cannot act as a nucleophile.
- At *high* pH (strongly alkaline conditions), there is not enough acid to protonate the OH group of the hemiaminal to make this into a better leaving group. The best compromise is around pH 4.5.

There are a number of related addition-elimination reactions, which employ similar nucleophiles.

8.3.7.2   *Reactions of imines, oximes and hydrazones*

The most important reaction of imines (e.g. $RCH=NR$) is their reduction to form amines (e.g. RCH_2-NHR). The conversion of an aldehyde or ketone into an amine, via an imine, is called *reductive amination*. This allows, for example, the selective formation of a secondary amine (R_2NH) from a primary amine (RNH_2).

Imines can also be attacked by nucleophiles, such as cyanide (NC^-), and this is used in the *Strecker amino acid synthesis*.

The mechanism of hydrolysis of a nitrile, RCN, is discussed in Section 9.9

The carbon atom next to a C=O bond is called the alpha (α) carbon (Section 8.4), hence the term α-amino acid

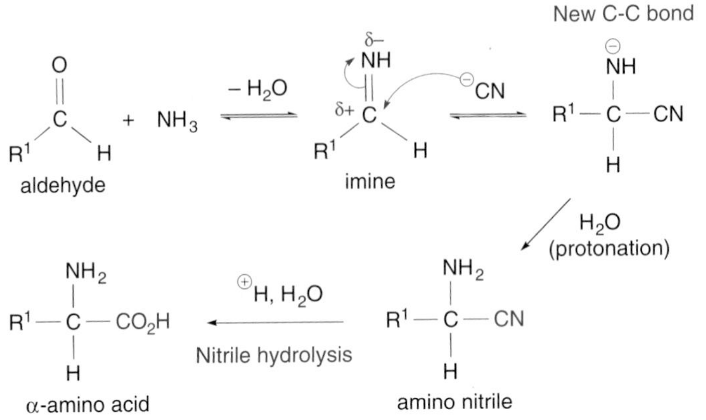
New C-C bond
− H_2O
aldehyde
imine
H_2O (protonation)
$^{\oplus}H$, H_2O
Nitrile hydrolysis
α-amino acid
amino nitrile

An important reaction of oximes (e.g. $R_2C=NOH$) is their conversion into amides (e.g. RCONHR) in the *Beckmann rearrangement reaction*.

Resonance is discussed in Section 1.6.3

For reactions of amides, see Section 9.8

Tautomerism is introduced in Section 8.4.1

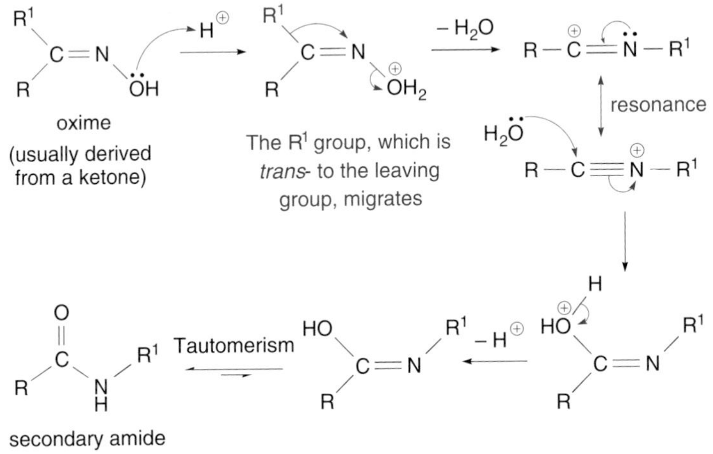
oxime
(usually derived from a ketone)
The R^1 group, which is *trans*- to the leaving group, migrates
− H_2O
resonance
Tautomerism
− $H^{\oplus}$
secondary amide

An important reaction of hydrazones (e.g. $R_2C=N\text{–}NH_2$) is their conversion into alkanes (e.g. R_2CH_2) on heating with hydroxide. This is called the *Wolff-Kischner reaction*.

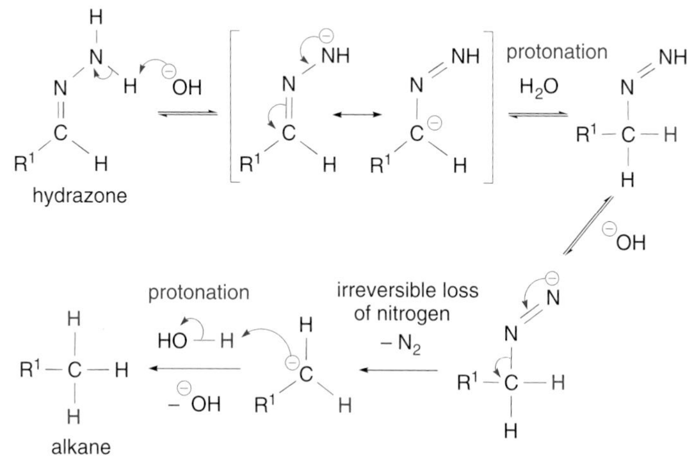

The hydrazone shown in the scheme is derived from an aldehyde and hydrazine

hydrazone

Nitrogen gas is an excellent leaving group (N_2 has a strong triple bond and as it is a gas, the reaction is favoured on entropic grounds, Section 4.9.1.2)

protonation — irreversible loss of nitrogen — $-N_2$

alkane

An aldehyde (RCHO) or ketone (RCOR) can therefore be reduced to an alkane via a hydrazone. This transformation can also be achieved using the *Mozingo reduction* (see Section 8.3.6) or the *Clemmensen reduction* (shown below; see Section 7.6).

Oxidation levels can be used to assess if a reaction involves reduction (Section 4.5)

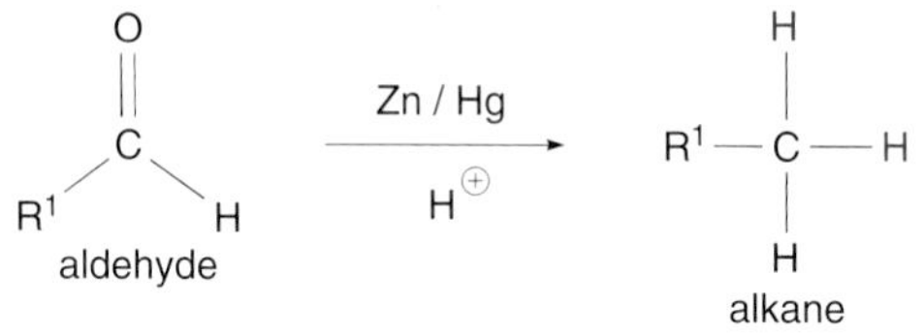

aldehyde   alkane

8.3.7.3 Formation of enamines

Enamines are prepared from reaction of an aldehyde (RCHO) or ketone (RCOR) with a *secondary* amine (R_2NH).

The carbon atom next to a C=O bond is called the alpha (α) carbon atom (Section 8.4)

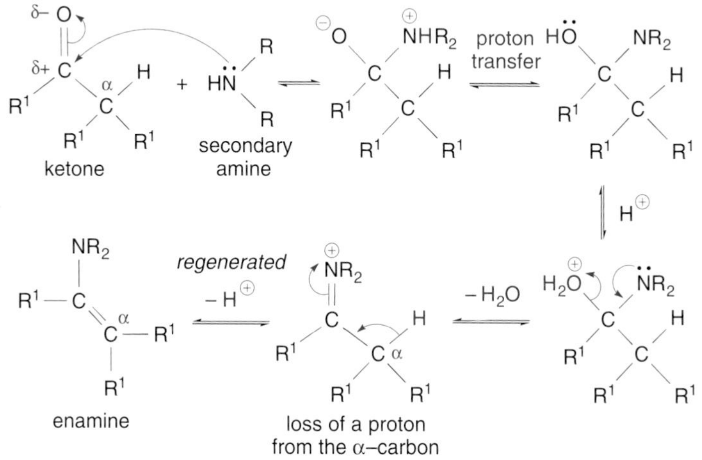

ketone   secondary amine   proton transfer

Notice that an acid catalyst is required to convert the OH group into a good leaving group

enamine   regenerated   loss of a proton from the α–carbon

The name enamine is derived from a combination of alkene and amine

8.4 α-Substitution reactions

The name enol is derived from a combination of alkene and alcohol

These reactions take place at the position next to the carbonyl group (i.e. the α-position) and involve substitution of an α-hydrogen atom by another group. The reactions take place via *enol* or *enolate ion* intermediates.

8.4.1 Keto-enol tautomerism

Catalysis is introduced in Section 4.9.2

Carbonyl (or keto) compounds (e.g. $RCOCH_3$) are interconvertible with their corresponding enols (e.g. $RC(OH)=CH_2$). This rapid interconversion of structural isomers under ordinary conditions is called *tautomerism*. Keto-enol tautomerism is catalysed by acids or bases.

Deprotonation of an α-hydrogen atom forms an enolate ion that is stabilised by resonance (Section 1.6.3)

Tautomers are compounds that are interconverted by tautomerism

Conjugation is introduced in Section 1.6.3

This is an example of *prototropy*, which is the movement of an (acidic) hydrogen atom and a double bond.

Hydrogen bonds are introduced in Section 1.1

For most carbonyl compounds, the keto structure is greatly preferred, mainly due to the extra strength of the C=O bond (a C=O bond is stronger than a C=C bond). However, the enol form is stabilised if the C=C bond is *conjugated* with a second π system or if the OH group is involved in *intramolecular hydrogen bonding*. Example

Notice that the intramolecular H-bond forms a 6-membered ring

The amounts of keto and enol forms at equilibrium is affected by the solvent; the enol form is more stable in a nonpolar solvent

8.4.2 Reactivity of enols

Enols (e.g. $RC(OH) = CR_2$) behave as nucleophiles and react with electrophiles at the α-position.

The name enol is derived from a combination of alkene and alcohol

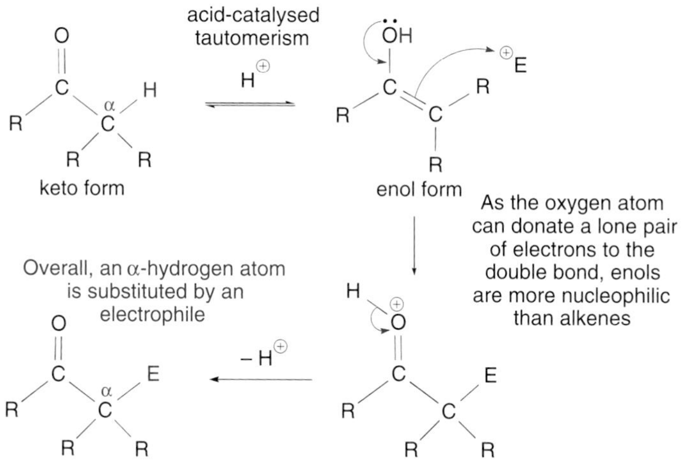
acid-catalysed tautomerism
keto form
enol form
As the oxygen atom can donate a lone pair of electrons to the double bond, enols are more nucleophilic than alkenes
Overall, an α-hydrogen atom is substituted by an electrophile

Catalysis is introduced in Section 4.9.2

Enols could react with electrophiles (E) to form a new O–E or C–E bond. As shown in the scheme, most electrophiles react to form a new C–E bond

8.4.2.1 *α-Halogenation of aldehydes and ketones*

Halogenation can be achieved by reaction of an aldehyde or ketone with chlorine, bromine or iodine in acidic solution.

Example

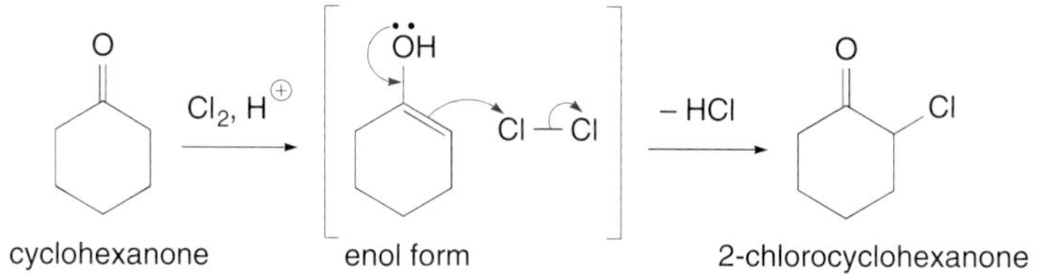
cyclohexanone
enol form
2-chlorocyclohexanone

On approaching the C=C bond the halogen becomes polarised (Section 6.2.2.2)

For naming carbonyl compounds, see Section 2.4

8.4.3 Acidity of α-hydrogen atoms: enolate ion formation

Bases can abstract α-hydrogen atoms from carbonyl compounds to form enolate ions.

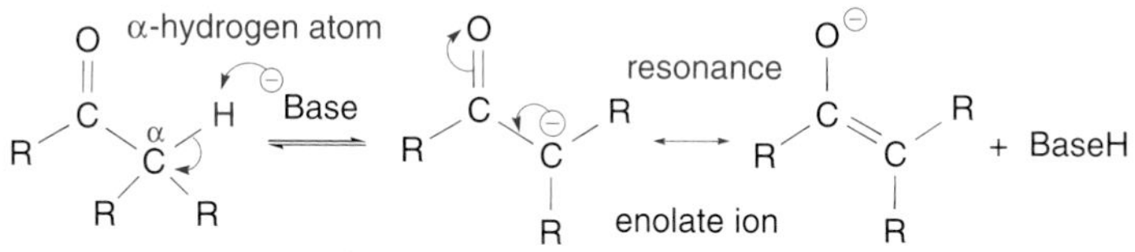
α-hydrogen atom
Base
resonance
enolate ion
+ BaseH

Bases are introduced in Section 1.7.2

Carbonyl compounds are more acidic than, for example, alkanes because the anion is stabilised by resonance (see Section 4.3.1).

Resonance stabilisation of carbanions is discussed in Section 1.6.3

Acidity is introduced in Section 1.7.1

pK_a	Compound	Anion
60	H_3C-CH_3	$H_3C-CH_2^{\ominus}$
20	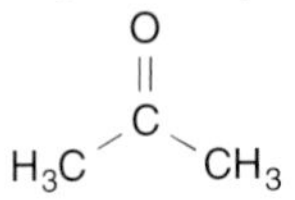 (acetone)	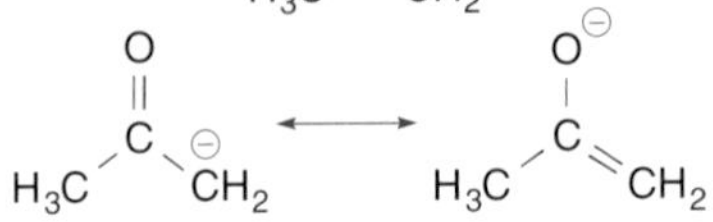 (enolate)
9	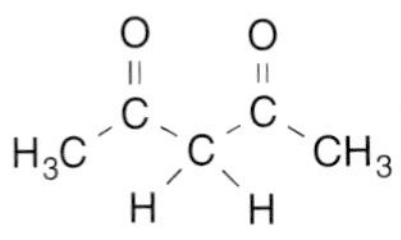 (1,3-diketone)	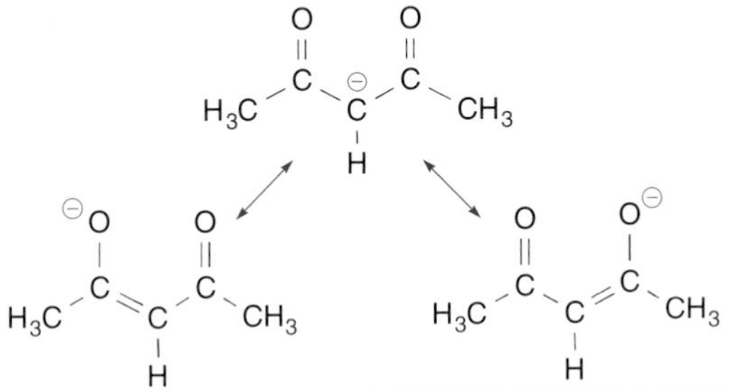

1,3-Diketones (or β-diketones) are therefore more acidic than water (which has a pK_a value of 16), as the enolate ion is stabilised by resonance over both carbonyl groups.

Nucleophiles, like enolate ions, that can react with electrophiles at two or more sites are called ambident nucleophiles

8.4.4 Reactivity of enolates

Enolate ions are much more reactive towards electrophiles than enols because they are negatively charged. Enolate ions can react with electrophiles on oxygen although reaction on carbon is more commonly observed.

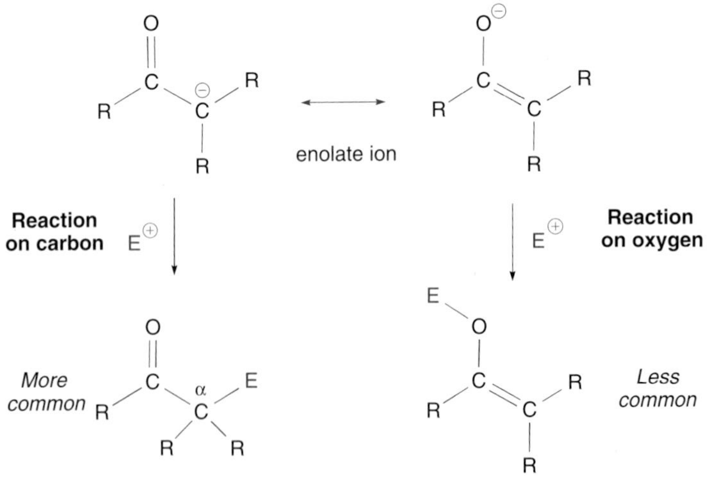

enolate ion

Reaction on carbon E⊕

Reaction on oxygen E⊕

More common

Less common

8.4.4.1   Halogenation of enolates

Reaction of methyl ketones (RCOCH$_3$) with excess hydroxide and chlorine, bromine or iodine leads to a carboxylic acid (RCO$_2$H) together with CHX$_3$ in a *haloform reaction*. The use of iodine gives CHI$_3$ (iodoform or triiodomethane), which is the basis of a functional group test for methyl ketones.

Resonance stabilisation of carbanions is discussed in Section 1.6.3

Acidity is introduced in Section 1.7.1

The 3 electronegative iodine atoms can help to stabilise the charge

The electron-withdrawing Cl$_3$ group increases the δ+ character of the carbonyl carbon atom

Therefore, whereas halogenation under acidic conditions leads to a monohalogenated product (e.g. RCOCH$_2$X), halogenation under basic conditions leads to a polyhalogenated product.

8.4.4.2   Alkylation of enolate ions

S$_N$2 reactions of halogenoalkanes are discussed in Section 5.3.1.1

Enolate ions can be alkylated by reaction with halogenoalkanes, RX (in an S$_N$2 reaction with primary and secondary halogenoalkanes, RCH$_2$X and R$_2$CHX, respectively). These reactions produce new carbon–carbon bonds.

The common name for pentane-2,4-dione, CH$_3$COCH$_2$COCH$_3$, is acetylacetone

8.5 Carbonyl-carbonyl condensation reactions

These reactions, which involve both nucleophilic addition and α-substitution steps, are amongst the most useful carbon–carbon bond forming reactions in synthesis.

8.5.1 Condensations of aldehydes and ketones: the aldol condensation reaction

The aldol reaction is a base-catalysed dimerisation reaction for all aldehydes (e.g. RCH_2CHO) and ketones (e.g. $RCOCH_3$) with α-hydrogens.

Notice that all three steps are reversible (for a discussion of equilibria, see Section 4.9.1.1)

α-Deprotonation   Nucleophilic addition

ethanal   enolate ion

Reaction of propanone (CH_3COCH_3), under the same conditions, forms $CH_3COCH_2C(OH)(CH_3)_2$

H_2O
(protonation)

3-hydroxybutanal or aldol
(a β-hydroxyaldehyde)

New C–C bond

Elimination reactions are discussed in Section 5.3.2

With aldehydes (e.g. CH_3CHO), this rapid and reversible reaction leads to the formation of a β-hydroxy aldehyde or *aldol* (*ald* for aldehyde, *ol* for alcohol) (e.g. $CH_3CH(OH)CH_2CHO$). When using ketones (e.g. CH_3COCH_3), β-hydroxy ketones are formed (e.g. $CH_3COCH_2C(OH)(CH_3)_2$). The products of the aldol reaction can undergo loss of water on heating, under basic or acidic conditions, to form conjugated enals or enones (containing both C=O and C=C bonds, e.g.

A conjugated compound has alternating single and double bonds (Section 1.6.3)

$CH_3CH=CHCHO$ or $CH_3CH=CHCOCH_3$) in condensation reactions. Conjugation stabilises the enal or enone product and this makes it relatively easy to form.

Base catalysed elimination of water (E1cB mechanism)

β-hydroxy ketone   enolate ion   enone
(*E*-isomer)

C=O and C=C bonds
are conjugated

The E1cb reaction is discussed in Section 5.3.2.4

The more stable C=C bond, with the *E*-configuration, is formed

Acid catalysed elimination of water

keto-enol tautomerism

β-hydroxy ketone   enol   enone
(*E*-isomer)

C=O and C=C bonds
are conjugated

For assigning alkenes as *E* or *Z* configuration, see Section 3.3.1.2

Protonation of the OH group converts it into a good leaving group, namely water (Section 5.3.1.4)

8.5.2   Crossed or mixed aldol condensations

These reactions involve two *different* carbonyl partners. If two similar aldehydes or ketones are reacted, a mixture of all four possible products is likely and the yield of any one product is low (e.g. reaction of A and B gives AA, AB, BA and BB). This is because both carbonyl partners can act as nucleophiles and electrophiles.

The formation of only one product requires the following:

Reaction of, for example, ethanal (CH₃CHO) and propanal (CH₃CH₂CHO), gives a mixture of four aldol products

1. Only one carbonyl partner can have α-hydrogens. This means that only one carbonyl partner can be deprotonated to form an enolate ion nucleophile.
2. The carbonyl partner without α-hydrogens must be more electrophilic than the carbonyl partner with α-hydrogens.
3. The carbonyl partner with α-hydrogen atoms should be added slowly to the reaction mixture. This will ensure that as soon as the enolate ion is generated it will be trapped by the carbonyl without α-hydrogen atoms (as this is present in high concentration).

α-Hydrogens are hydrogen atoms on a carbon atom adjacent to a C=O group

Example

'Bu stands for *tertiary*-butyl and is an abbreviation for –C(CH$_3$)$_3$; Ph stands for phenyl and is an abbreviation for –C$_6$H$_5$ (Section 2.2)

The carbon atom in the C=O bond of an aldehyde is more electrophilic than the carbon atom in the C=O bond of a ketone (Section 8.3.1)

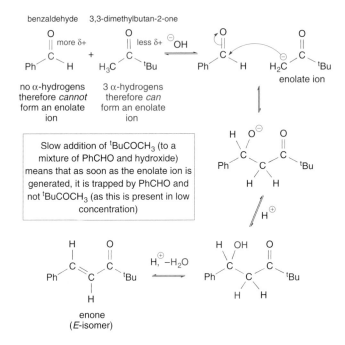

benzaldehyde   3,3-dimethylbutan-2-one

no α-hydrogens therefore *cannot* form an enolate ion

3 α-hydrogens therefore *can* form an enolate ion

enolate ion

Slow addition of 'BuCOCH$_3$ (to a mixture of PhCHO and hydroxide) means that as soon as the enolate ion is generated, it is trapped by PhCHO and not 'BuCOCH$_3$ (as this is present in low concentration)

enone
(*E*-isomer)

8.5.3   Intramolecular aldol reactions

In an intramolecular reaction, different parts of the same molecule react. If a ring is formed, this is called a cyclisation reaction

Intramolecular cyclisation reactions can occur so as to form stable 5- or 6-membered cyclic enones. These ring sizes are preferred over strained 3- and 4-membered rings (see Section 3.2.3) or difficult to make medium sized (8–13-membered) rings. (Large rings are difficult to make because as the precursor chain length increases, so does the number of conformations that can be adopted. This means that, for long chains, there is a greater loss of entropy on formation of the cyclisation transition-state.)

Example

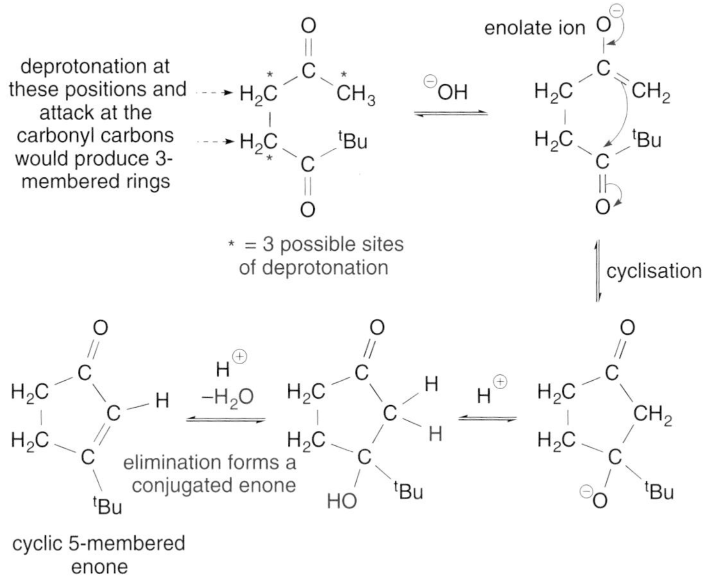

deprotonation at
these positions and
attack at the
carbonyl carbons
would produce 3-
membered rings

* = 3 possible sites
of deprotonation

cyclisation

elimination forms a
conjugated enone

cyclic 5-membered
enone

Notice that all of the steps are
reversible

A conjugated compound has
alternating single and double
bonds, e.g. O=C–C=C
(Section 1.6.3)

The product is a substituted
cyclopent-2-enone (Section 2.2)

8.5.4 The Michael reaction

Enones (e.g. $H_2C = CH–COCH_3$) derived from aldol condensation reactions can
undergo further carbon–carbon bond forming reactions on addition of carbon
nucleophiles. When the nucleophile adds at the 4-position (rather than the 2-
position) of the enone, this is called the *Michael reaction* (or 1,4-addition or
conjugate addition).

1,2-Addition

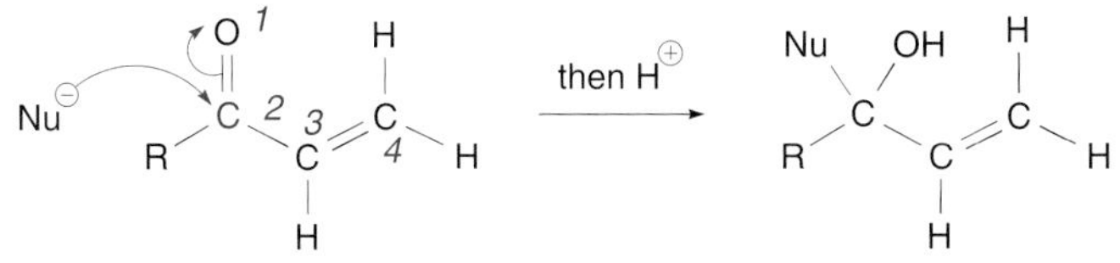

then $H^{\oplus}$

attack at the carbonyl carbon

Notice that in both 1,2- and 1,4-
addition, a pair of electrons moves
on to the oxygen atom of the enone

The enone in a Michael reaction is
sometimes called the Michael
acceptor

1,4-Addition

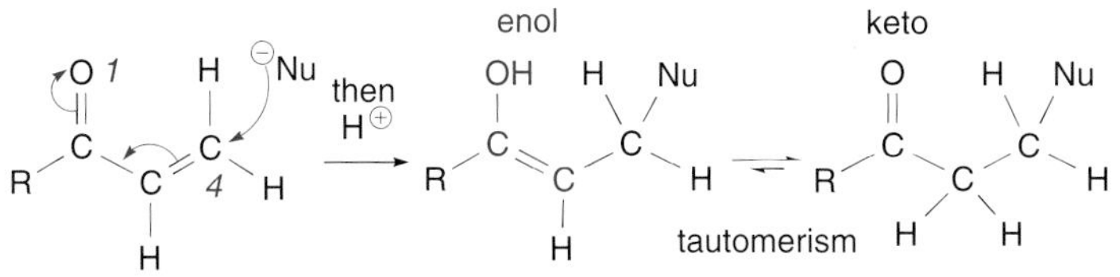

then
$H^{\oplus}$

enol   keto

tautomerism

attack at the conjugated alkene

Keto-enol tautomerism is discussed
in Section 8.4.1

Organometallics are introduced in Section 8.3.4.2

The site of attack can depend on the nature of the nucleophile. Organolithium compounds (RLi) and Grignard reagents (RMgX) tend to give 1,2-addition. For 1,4-addition, organocopper reagents (e.g. RCu, or R_2CuLi, which are called cuprates) can be used. The change in site of attack has been explained by the *hard and soft principle*, as described below.

RCu can be formed from Li and CuI; R_2CuLi can be formed from RCu and RLi

A large delocalised nucleophile, with the charge spread out, such as MeCOCH⁻COMe, reacts by 1,4-addition, rather than 1,2-addition

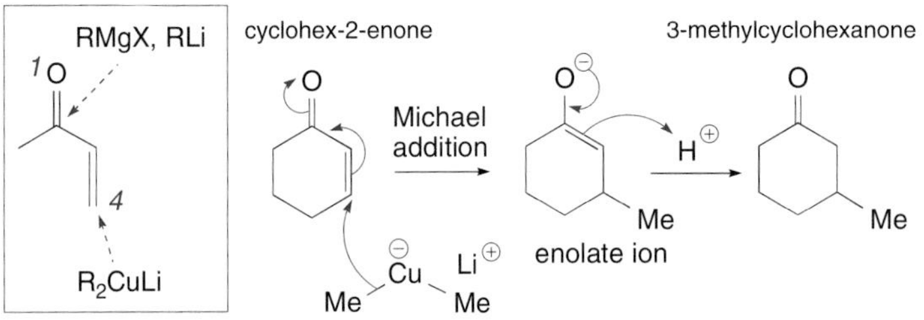

- *Hard nucleophiles* are those with a negative charge localised on one small atom (e.g. methyllithium, MeLi). These tend to react at the carbonyl carbon, which is known as a *hard electrophilic centre* (it has a high δ+ density as it is directly bonded to oxygen).
- *Soft nucleophiles* are those in which the negative charge is delocalised (e.g. spread over the large copper atom in R_2CuLi). These tend to react at the 4-position, which is called a *soft electrophilic centre* (it has a low δ+ density).

Worked example

The following questions are based on the reactions of propanal **1**, shown below

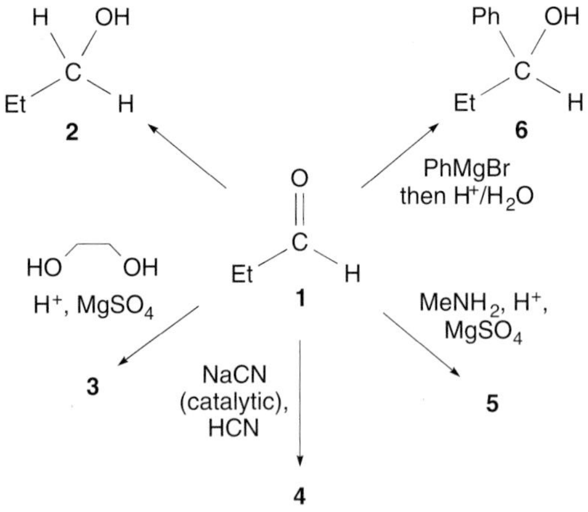

Hint: Consider oxidation levels (Section 4.5)

(a) Give the reagents needed to convert **1** into **2**. Is this an oxidation or reduction reaction?

(b) Give the structures of compounds **3**, **4** and **5**, and name the functional groups that are present in these compounds.

Hint: Consider nucleophilic addition reactions

(c) The rate of formation of **5** depends on the pH of the reaction mixture. Why is the reaction fastest at about pH 4.5?

Hint: Consider the mechanism of formation of an imine

(d) Provide a reaction mechanism to show **1** is converted into **6**.

PhMgBr is a strong nucleophile

Answer

(a) NaBH$_4$ or LiAlH$_4$ then H$^+$. Reduction.

For reaction of complex metal hydrides, see Section 8.3.3.1

(b) (cyclic) acetal   cyanohydrin   imine

For addition of cyanide, see Section 8.3.4.1; for addition of alcohols, see Section 8.3.5.2; for addition of primary amines, see Section 8.3.7.1

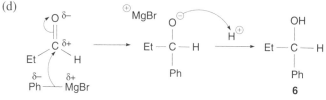

(c) At low pH the amine (MeNH$_2$) is protonated and so cannot act as a nucleophile. At high pH there is not enough acid to protonate the OH group of the hemiaminal to convert it into a good leaving group.

For imine formation, see Section 8.3.7.1

(d)

For addition of Grignard reagents, see Section 8.3.4.2

Problems

1. The organic product of the reaction of an aldehyde (RCHO) with NaBH$_4$ followed by aqueous acid is RCH$_2$OH.

   (a) Provide a mechanism for this reaction.

   Section 8.3.3

   (b) Suggest a method for conversion of RCH$_2$OH back to RCHO.

   Section 8.3.3.5

2. The organic product from the reaction of methanal with phenylmagnesium bromide, PhMgBr, followed by aqueous acid, is PhCH$_2$OH.

   Section 8.3.4.2

   (a) Provide a mechanism for this reaction.

   (b) Suggest how you would prepare PhMgBr.

3. Ethanoylbenzene (acetophenone), PhCOCH$_3$, reacts with iodine-potassium iodide solution and aqueous sodium hydroxide to give a yellow solid **A**, relative molecular mass 394, which is removed on filtration. Treatment of the aqueous filtrate with HCl produces a white precipitate of compound **B**, which shows an absorption band at 1700 cm^{-1} in the infrared spectrum.

   Section 8.4.4.1

   Give the structures and outline the mechanism of formation of compounds **A** and **B**.

Section 8.3.5.2

4. Cyclopentanone reacts with ethane-1,2-diol ($HOCH_2CH_2OH$) in the presence of acid to form a compound **C**, $C_7H_{12}O_2$.

 (a) Give a structure for **C**.
 (b) Explain the role of the acid.
 (c) Provide a mechanism for the reaction.

Section 8.5.1

5. Reaction of CH_3CHO with aqueous sodium hydroxide produces a new compound **D**, $C_4H_8O_2$. When **D** is heated with dilute acid, compound **E** is formed together with water.

 (a) Give a structure for **D** and provide a mechanism for its formation.
 (b) Give a structure for **E** and provide a mechanism for its formation.

Section 8.3.7.2

6. Reaction of 3-methyl-1-butanol with pyridinium chlorochromate (PCC) gives compound **F**. When **F** is treated with ammonium chloride and potassium cyanide a new compound, **G**, is formed. Heating **G** with aqueous hydrochloric acid then affords a salt **H**, $C_6H_{13}NO_2 \cdot HCl$.

 Give structures for compounds **F** – **H**.

7. The following questions relate to the reaction scheme shown below.

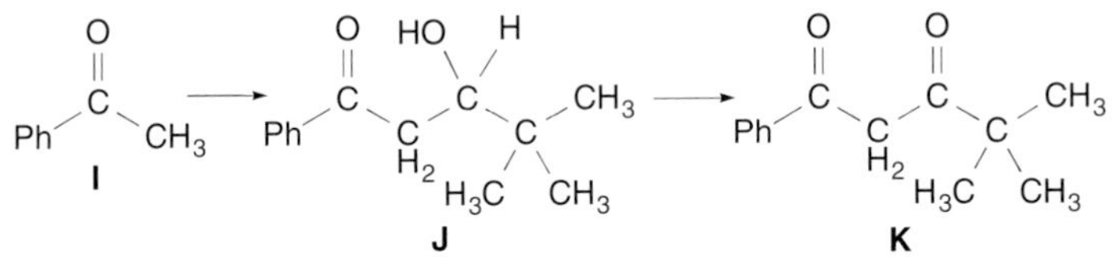

I → **J** → **K**

Section 8.5.1
(a) Suggest reagents to convert **I** into **J**.

Section 8.5.1
(b) Heating **J** with acid produces a new compound with the formula $C_{13}H_{16}O$. Draw the structure of this new compound, indicating any stereochemistry.

Section 8.3.3.5
(c) Suggest reagents to convert **J** into **K**.

Section 8.4.1
(d) Giving your reasons, draw the most stable enol form of compound **K**.

9

Carbonyl compounds: carboxylic acids and derivatives

Key point. Carboxylic acids (RCO_2H) and carboxylic acid derivatives, such as esters and amides, have an electronegative group (e.g. OH, OR, NHR or halogen) directly bonded to the carbonyl (C=O). These compounds generally undergo *nucleophilic acyl substitution reactions* in which a nucleophile replaces the electronegative group on the carbonyl. The inductive and mesomeric effects of the electronegative group determine the relative reactivity of these compounds towards nucleophilic attack. Carboxylic acid derivatives with an α-hydrogen atom can form enolate ions (on deprotonation), which can react in *α-substitution reactions* or in *carbonyl-carbonyl condensation reactions*.

nucleophilic acyl substitution   *ester enolate ion*

9.1 Structure

Carboxylic acids (RCO_2H) are members of a class of acyl compounds which all contain an electronegative group bonded to RCO.

Y	Functional Group
OH	carboxylic acid
halogen	acid (acyl) halide
OCOR	acid anhydride
OR	ester
NH$_2$, NHR, NR$_2$	amide

Functional groups are introduced in Section 2.1

RCONH$_2$ is a primary amide, RCONHR is a secondary amide, and RCONR$_2$ is a tertiary amide

Keynotes in Organic Chemistry, Second Edition. Andrew F. Parsons.
© 2014 John Wiley & Sons, Ltd. Published 2014 by John Wiley & Sons, Ltd.

9.2 Reactivity

Leaving groups are discussed in Section 5.3.1.4

As the electronegative group (Y) can act as a leaving group, carboxylic acid derivatives undergo nucleophilic acyl substitution reactions (i.e. reactions leading to substitution of the Y group by the nucleophile).

With charged nucleophiles

Sometimes an abbreviated mechanism is drawn, where the intermediate alkoxide ion is not shown

All steps are reversible; for the equilibrium to lie to the product, Y^- must be more stable than Nu^-

Y = halogen, OCOR, OR or NH_2 (NHR, NR_2). Nucleophiles = $H^{\ominus}$, $R^{\ominus}$

With uncharged nucleophiles

Y = halogen, OCOR, OR or NH_2 (NHR, NR_2)
Nucleophiles = H_2O, ROH, NH_3, RNH_2, R_2NH
This process can be acid catalysed

For reactions of aldehydes and ketones, see Sections 8.4 and 8.5

Carboxylic acid derivatives, like aldehydes (RCHO) and ketones (RCOR), can also undergo α-substitution and carbonyl-carbonyl condensation reactions.

9.3 Nucleophilic acyl substitution reactions

9.3.1 Relative reactivity of carboxylic acid derivatives

Rate-determining steps are introduced in Section 4.9.2.1

The addition of the nucleophile to the carbonyl carbon is usually the rate-determining step in the substitution reaction. Therefore the more electropositive the carbon atom (of the acyl group), the more reactive the carboxylic acid derivative.

most reactive   *least reactive*

Acid chloride   *Acid*   *Ester*   *Amide*
(Acyl chloride)   *anhydride*

The difference in reactivity is explained by comparing the inductive and mesomeric effects of the electronegative substituent (Y, in RCOY). For example, the $+M$ effect of Cl is much weaker than that of the NH_2, NHR or NR_2 group because the lone pair on chlorine does not interact well with the 2p orbital on carbon (see Section 7.3.2.2). Therefore, whereas Cl withdraws electrons from the carbonyl group, the NH_2 group mesomerically donates electrons towards the carbonyl group.

For inductive effects, see Section 1.6.1; for mesomeric effects, see Section 1.6.3

For Cl: $-I > +M$
The chlorine *withdraws* electrons from the acyl group making the carbonyl carbon very δ+ and so very susceptible to nucleophilic attack

For NH_2: $+M > -I$
The nitrogen *donates* electrons to the acyl group making the carbonyl carbon less δ+ and so less susceptible to nucleophilic attack

For an ester, the $+M$ of the OR group is also stronger than the $-I$ effect. The $+M$ effect of the OR group is not as strong as the $+M$ effect of an NH_2 (or NHR or NR_2) group of an amide

It is usually possible to transform a more reactive carboxylic acid derivative into a less reactive derivative (e.g. an acid chloride, RCOCl, can be converted into an amide, $RCONH_2$).

9.3.2   Reactivity of carboxylic acid derivatives versus carboxylic acids

As a carboxylic acid (RCO_2H) has an acidic hydrogen atom, nucleophiles may act as bases and deprotonate the acid rather than attack the carbonyl carbon atom.

The acidity of carboxylic acids, together with the influence of inductive effects, is discussed in Section 1.7.1

Nucleophiles attack the δ+ carbon

ester

Nucleophiles can deprotonate the acidic hydrogen rather than attack the δ+ carbon

carboxylic acid

9.3.3   Reactivity of carboxylic acid derivatives versus aldehydes/ketones

Aldehydes (RCHO) and ketones (RCOR) are generally *more* reactive to nucleophiles than esters (RCO_2R) or amides (e.g. $RCONR_2$) because the carbonyl carbon atom is more electropositive. In esters, the $+M$ effect of the OR group means that electrons are donated to the acyl group, lowering its reactivity to nucleophiles.

Reactions of esters are discussed in Section 9.7

most reactive   *least reactive*

acid anhydride   aldehyde   ester

In symmetrical anhydrides both R groups are the same; mixed anhydrides have different R groups

Normally, mesomeric effects are
stronger than inductive effects
(Section 1.6.3)

Aldehydes and ketones are generally *less* reactive to nucleophiles than acid halides (RCOCl) or acid anhydrides (RCO$_2$COR) because the carbonyl carbon atom is less electropositive. In acid anhydrides the carbonyl carbon is more electropositive than in an aldehyde because the oxygen lone pair (of the OCOR group) is shared between two carbonyl groups and so the +M effect is significantly weakened, hence −I > +M.

9.4 Nucleophilic substitution reactions of carboxylic acids

Good leaving groups are neutral
molecules or stable anions

Nucleophilic substitution of the OH group of a carboxylic acid (RCO$_2$H) is difficult to achieve because of competing deprotonation (see Section 9.3.2). In order for substitution to occur, the OH group needs to be converted into a good leaving group (e.g. Cl).

9.4.1 Preparation of acid chlorides

Thionyl chloride also reacts with
alcohols to form chloroalkanes
(Section 5.2.2)

Carboxylic acids (RCO$_2$H) can be converted into acid chlorides (acyl chlorides) using thionyl chloride (SOCl$_2$) or phosphorus trichloride (PCl$_3$).

S$_N$2 reactions are discussed in
Section 5.3.1.1

Resonance stabilisation of cations
is discussed in Section 1.6.3

For other Ei reactions, see Section
6.2.1

The Ei reaction leads to an increase
in entropy (Section 4.9.1.2)

9.4.2 Preparation of esters (esterification)

Catalysts are discussed in
Section 4.9.2

Carboxylic acids (RCO$_2$H) can be converted into esters (RCO$_2$R) by reaction with an alcohol (ROH) in the presence of an acid catalyst.

The text beside the first reaction scheme reads:

Protonation of the C=O oxygen of the carboxylic acid forms a more stable cation than that formed on protonation of the OH group

H^+ is rapidly exchanged between OR and OH groups

The reaction is reversible and the formation of the ester usually requires an excess of the alcohol.

9.5   Nucleophilic substitution reactions of acid chlorides

Acid chlorides (acyl chlorides), RCOCl, are very reactive and can be converted into a variety of compounds including less reactive carboxylic acid derivatives.

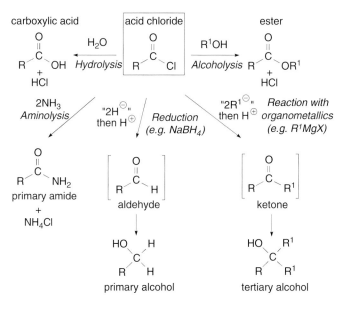

For the formation of Grignard reagents, RMgX, see Section 8.3.4.2

For the mechanism of reduction of aldehydes and ketones using $NaBH_4$, see Section 8.3.3.1

For the mechanism of reaction of aldehydes and ketones with Grignard reagents see Section 8.3.4.2

In all cases, nucleophilic acyl substitution leads to the introduction of a nucleophile at the expense of chlorine. Reaction with reducing agents or organometallic compounds (e.g. RMgX) can lead to intermediate aldehydes (RCHO) or ketones (RCOR). These can subsequently undergo nucleophilic addition reactions (to give alcohols) in the presence of a second equivalent of the reducing agent or organometallic compound.

Example

Notice that deprotonation of the tetrahedral intermediate occurs before loss of Cl⁻

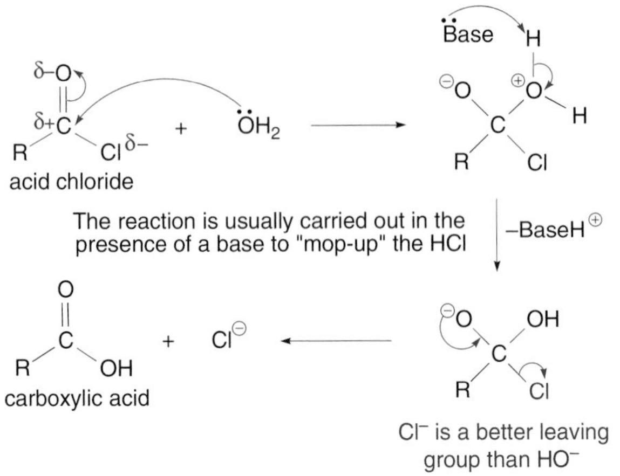

acid chloride

The reaction is usually carried out in the presence of a base to "mop-up" the HCl

carboxylic acid

Cl⁻ is a better leaving group than HO⁻

Under the same conditions, carboxylic acids can undergo acid-base reactions to produce salts.

In these reactions, ammonia (NH_3) and the Grignard reagent (RMgX) act as bases

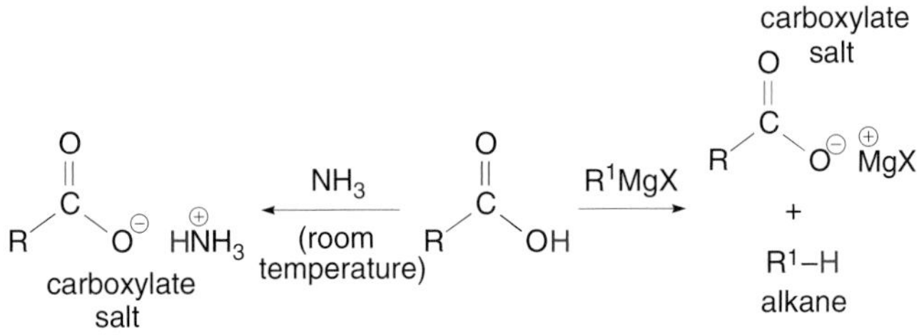

9.6 Nucleophilic substitution reactions of acid anhydrides

Acid anhydrides (RCO_2COR) undergo similar reactions to acid chlorides. These reactions involve the substitution of a carboxylate group ($RCO_2{}^-$). Reduction using $LiAlH_4$ produces an intermediate aldehyde (RCHO) and carboxylate ion ($RCO_2{}^-$), which are subsequently reduced into primary alcohols (RCH_2OH).

Acid anhydrides can be formed by dehydration of carboxylic acids using a dehydrating agent, such as phosphorus pentoxide (P_4O_{10})

For the reaction with ammonia (NH_3), one equivalent acts as a nucleophile and the second equivalent acts as a base

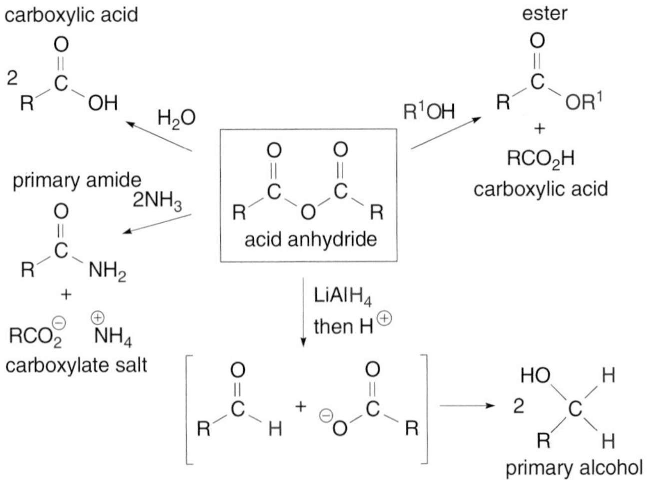

LiAlH$_4$ (but not NaBH$_4$) can reduce carboxylic acids (RCO$_2$H) into primary alcohols (RCH$_2$OH) by the mechanism shown below.

LiAlH$_4$ is a stronger reducing agent than NaBH$_4$ (Section 8.3.3.1)

The carbon atom in RCO$_2$H has an oxidation level of 3, while the carbon atom in RCH$_2$OH has an oxidation level of 1 (Section 4.5)

A stronger O–Al bond is formed at the expense of a weaker H–Al bond

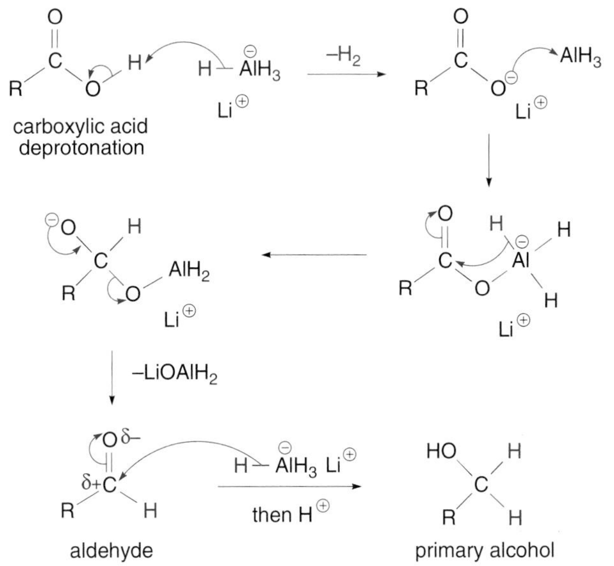

carboxylic acid deprotonation

−LiOAlH$_2$

aldehyde   then H$^{\oplus}$ primary alcohol

A particularly good reagent for reducing carboxylic acids into primary alcohols is borane (B$_2$H$_6$, which acts as BH$_3$). For a related reduction, see Section 9.8.

9.7 Nucleophilic substitution reactions of esters

Although less reactive towards nucleophiles than acid chlorides (RCOCl) or anhydrides (RCO$_2$COR), esters (RCO$_2$R) can react with a number of nucleophiles. Reduction usually requires LiAlH$_4$ as NaBH$_4$ reacts very slowly with esters. Indeed, NaBH$_4$ can be used to selectively reduce aldehydes (RCHO) and ketones (RCOR) in the presence of esters.

LiAlH$_4$ is a stronger reducing agent than NaBH$_4$ (Section 8.3.3.1)

carboxylic acid

+ R^{1}OH

H$_2$O, H$^{\oplus}$ or H$_2$O, HO$^{\ominus}$ then H$^{\oplus}$

ester

2"R$^{\ominus}$" then H$^{\oplus}$

tertiary alcohol

(e.g. RMgX)

primary amide

NH$_3$

+ R^{1}OH

"H$^{\ominus}$" then H$^{\oplus}$ (Using LiAlH$_4$)

primary alcohol
+ R^{1}OH

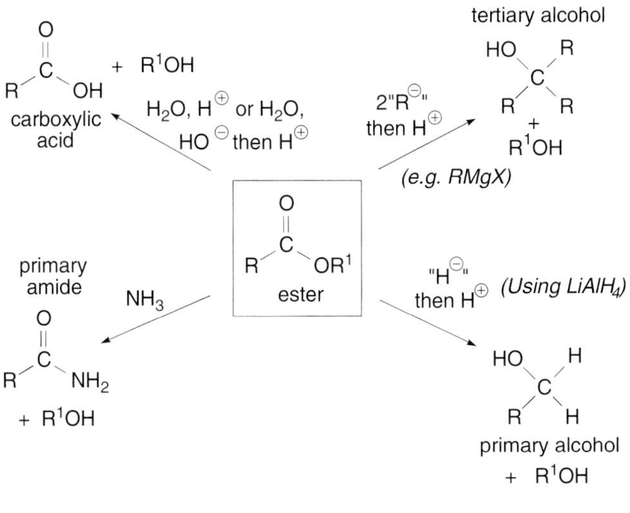

Esters react with two equivalents of Grignard reagents (RMgX) to form tertiary alcohols (R₃COH). This involves a nucleophilic substitution reaction (to form an intermediate ketone, RCOR) followed by a nucleophilic addition reaction.

For the mechanism of reaction of ketones with Grignard reagents, see Section 8.3.4.2

The R¹O⁻ group acts as a leaving group

Ketones are more reactive to nucleophiles than esters (Section 9.3.3)

Esters (RCO₂R) can be hydrolysed in aqueous acid (see Section 9.4.2) or base. In basic solution, this is known as a *saponification* reaction and this has an important application in the manufacture of soaps from vegetable oils or fats. Whereas acid hydrolysis is reversible, base hydrolysis is *irreversible* due to the formation of the resonance-stabilised carboxylate ion.

Resonance stabilisation of carboxylate ions (RCO₂⁻) is discussed in Section 1.7.1

Reaction of an ester (RCO₂R) with an alcohol (ROH) under basic or acidic conditions can produce a new ester. This is called a *transesterification* reaction. The equilibrium can be shifted to the product by removing the alcohol with a low boiling point from the reaction mixture using distillation.

9.8 Nucleophilic substitution and reduction reactions of amides

Amides ($RCONH_2$, RCONHR or $RCONHR_2$) are much less reactive than acid chlorides (RCOCl), acid anhydrides (RCO_2COR) or esters (RCO_2R). Harsh reaction conditions are required for cleavage of the amide bond while reduction requires $LiAlH_4$ or borane (B_2H_6, which reacts as BH_3).

In an amide, the nitrogen atom donates electrons to the acyl group making the carbonyl carbon less $\delta+$ (Section 9.3.1)

Reduction of primary amides ($RCONH_2$) forms primary amines (RCH_2NH_2), while secondary (RCONHR) and tertiary ($RCONR_2$) amides can be reduced into secondary (RCH_2NHR) and tertiary (RCH_2NR_2) amines, respectively. It should be noted that these reactions do not involve nucleophilic substitution of the NH_2 (or NHR or NR_2) group of the amide.

Lithium aluminium hydride reduction

Reactions of imines are discussed in Section 8.3.7.2

Borane reduction

BH$_3$ also reacts with alkenes
(Section 6.2.2.5)

A B$-$O bond is stronger than a
B$-$H bond

9.9 Nucleophilic addition reactions of nitriles

Nitriles (RCN) are related to carboxylic acids (RCO$_2$H) and derivatives in that the carbon atom of the nitrile is at the same oxidation level as the carbon atom of the acyl group. As a consequence, the reactions of nitriles are similar to that of carbonyls and nucleophiles attack the nitrile carbon atom.

Oxidation levels are introduced in
Section 4.5

The most important reactions of nitriles are hydrolysis, addition of organometallics (e.g. RMgX) and reduction. Reduction with lithium aluminium hydride (LiAlH$_4$) produces amines whereas diisobutylaluminium hydride (DIBAL-H) produces aldehydes. This is because DIBAL-H is a less powerful reducing agent than LiAlH$_4$ (in part because the reagent is more sterically hindered and therefore has more difficulty in transferring hydride ions). So whereas reaction with DIBAL$-$H stops at the imine, reaction with LiAlH$_4$ leads to further reduction of the imine (C$=$N bond) to give an amine.

DIBAL-H is an abbreviation for
HAl^{i}Bu$_2$

Steric hindrance is introduced in
Section 4.4

For hydrolysis of an imine (e.g. $R_2C{=}NH$) to form a ketone $(R_2C{=}O)$, see Section 8.3.7.1

Hydrolysis

Tautomerism is introduced in Section 8.4.1

Reduction with DIBAL-H

^{i}Bu stands for an isobutyl group $-CH_2CH(CH_3)_2$

For hydrolysis of an imine into an aldehyde, see Section 8.3.7.1

9.10 α-Substitution reactions of carboxylic acids

PBr₃ can also be used to convert an alcohol into a bromoalkane (Section 5.2.2)

Carboxylic acids (RCO_2H) can be brominated at the α-position by reaction with bromine and phosphorus tribromide (PBr_3) in the *Hell-Volhard-Zelinsky reaction*. The reaction, which proceeds via an acid bromide ($RCOBr$), leads to the substitution of an α-hydrogen atom by a bromine atom.

Keto-enol tautomerism is discussed in Section 8.4.1

The mechanism of hydrolysis of an acid bromide is analogous to the hydrolysis of an acid chloride (Section 9.5)

9.11 Carbonyl-carbonyl condensation reactions

For the formation of enolate ions from aldehydes and ketones, see Section 8.4.3

Esters (e.g. RCH_2CO_2R) with α-hydrogen atoms can be deprotonated (like aldehydes and ketones, RCH_2CHO and RCH_2COR) to form resonance-stabilised enolate ions, which can act as nucleophiles.

9.11.1 The Claisen condensation reaction

Equilibria is introduced in Section 4.9.1.1

A condensation reaction between two esters using one equivalent of base is called the *Claisen reaction*. One ester loses an α-H atom while the other loses an alkoxide ion (RO^-). The initial steps are reversible but deprotonation of the intermediate β-keto ester (by the alkoxide ion) shifts the equilibrium to the desired product. It should be noted that deprotonation of the β-keto ester forms an anion that can be stabilised by delocalisation over two carbonyl groups. At the end of the reaction, acid is added to reprotonate the condensation product.

In a condensation reaction, two molecules combine to give a product with loss of a small molecule, such as water or an alcohol

The negative charge in the β-keto ester enolate ion is stabilised by delocalisation over both C=O bonds

The alkoxide base and ester side chain should be matched. For example, the ethoxide ion (EtO$^-$) should be used as a base for ethyl esters (R = Et in the scheme above). This ensures that if the ethoxide ion attacks the carbonyl group of an ethyl ester (in a transesterification reaction, see Section 9.7), then an ethyl ester will be re-formed.

9.11.2 Crossed or mixed Claisen condensations

This involves a reaction between two different esters. The reaction works best when only one of the esters has α-hydrogen atoms (for the same reasons as for the crossed aldol reaction. See Section 8.5.2).

Example

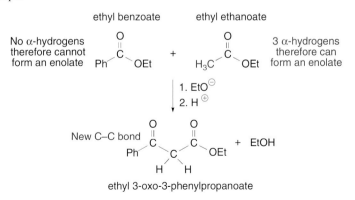

Et stands for ethyl (CH$_3$CH$_2$−); Ph stands for phenyl and is an abbreviation for −C$_6$H$_5$ (Section 2.2)

For naming esters, see Section 2.4

For the formation and reaction of ketone enolate ions, see Section 8.4.3

Crossed Claisen-like reactions can also occur between esters and ketones. The ester generally acts as the electrophile as ketones are more acidic than esters (i.e. the ketone enolate ion, which acts as the nucleophile, is more easily formed than an ester enolate ion). For an ester enolate ion, the lone pair of electrons on the OR group can compete with the negative charge for delocalisation onto the carbonyl group. This means that the negative charge of an ester enolate is not as readily delocalised onto the carbonyl group as that of a ketone enolate ion.

Resonance stabilisation of anions is discussed in Section 1.6.3

Acidity and pK_a values are introduced in Section 1.7.1

However, the reaction works best when the ester does not contain any α-hydrogen atoms (and so cannot form an ester enolate ion).

For naming carbonyl compounds, see Section 2.4

9.11.3 Intramolecular Claisen condensations: the Dieckmann reaction

For other intramolecular reactions, see Sections 8.3.5.2 and 8.5.3

Intramolecular Claisen reactions are called *Dieckmann reactions* and these work well to give 5- or 6-membered rings. A 1,6-diester forms a 5-membered ring while a 6-membered ring is formed from condensation of a 1,7-diester.

Example

Notice that the alkoxide ion and ester side chain are matched

9.12   A summary of carbonyl reactivity

The following guidelines can be used to help predict reaction pathways for aldehydes (RCHO), ketones (RCOR) and carboxylic acid derivatives.

Reactions of aldehydes and ketones are discussed in Chapter 8

(i) Nucleophilic addition versus nucleophilic substitution

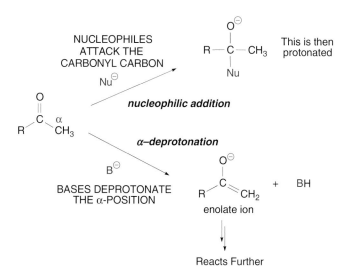

Leaving groups are introduced in Section 5.3.1.4

Addition and substitution reactions are introduced in Section 4.6

(ii) Nucleophilic addition versus α-deprotonation

Bases are discussed in Section 1.7.2

Worked example

The following questions are based on the reaction scheme below.

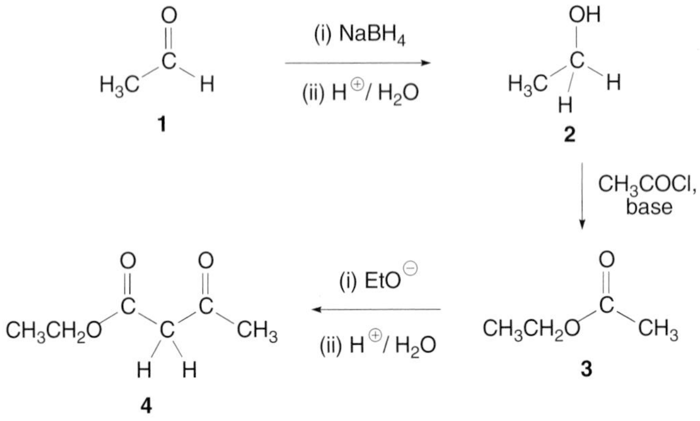

Hint: NaBH$_4$ is a source of H$^-$

(a) Provide a reaction mechanism (using curly arrows) to show how **1** is converted into **2**.

Hint: Consider ester hydrolysis

(b) Ester **3** can be converted back into alcohol **2** by heating with aqueous sodium hydroxide. Provide a mechanism (using curly arrows) for this saponification reaction.

Hint: Consider a Claisen condensation reaction

(c) Provide a mechanism (using curly arrows) for the conversion of ester **3** into β-keto ester **4**, which involves the intermediate formation of two different enolate ions.

Answer

For reduction of an aldehyde using a complex metal hydride, see Section 8.3.3.1

(a) NaBH$_4$ is a source of H$^-$.

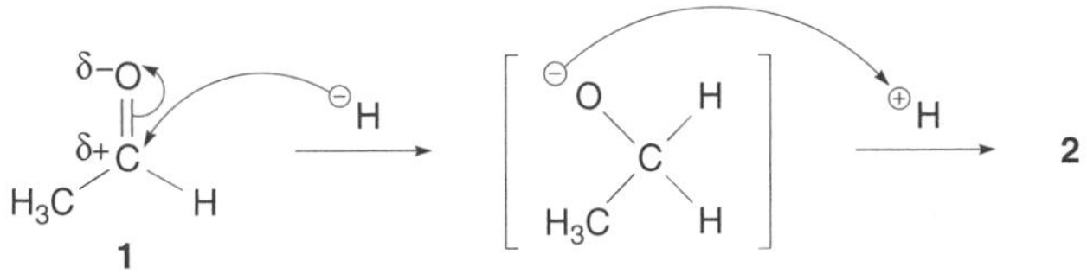

(b)

For ester hydrolysis under basic conditions, see Section 9.7

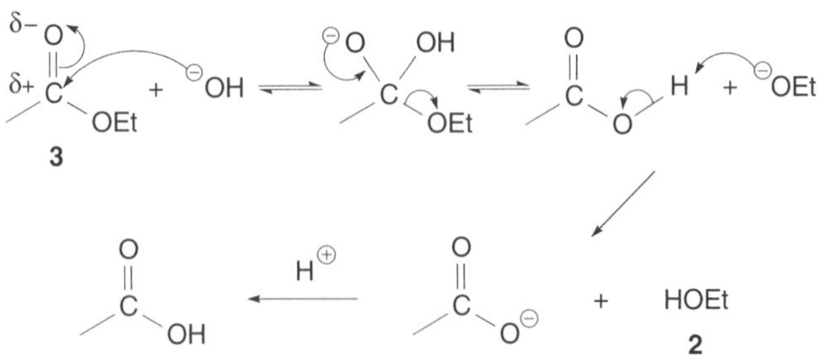

(c)

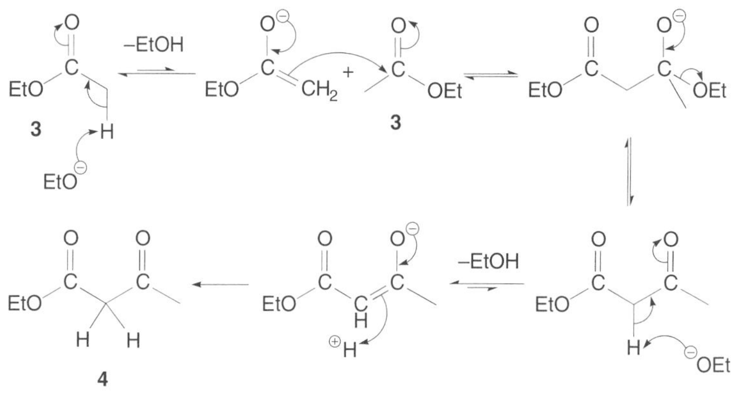

For the Claisen condensation reaction, see Section 9.11.1

Problems

1. Lactone **C** can be made from **A** by the following sequence of reactions.

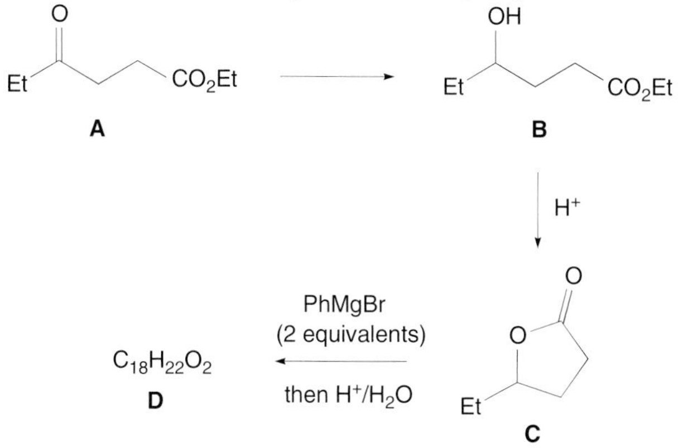

(a) How could **A** be converted into **B**? Explain why the ketone rather than the ester group reacts.

Sections 8.3.3.1 and 9.3.3

(b) Provide a mechanism to show how **B** can be converted into **C**.

Section 9.7

(c) Give the structure of product **D**.

Section 9.7

2. What reagents can be used to accomplish the following transformations?

Sections 9.4–9.9

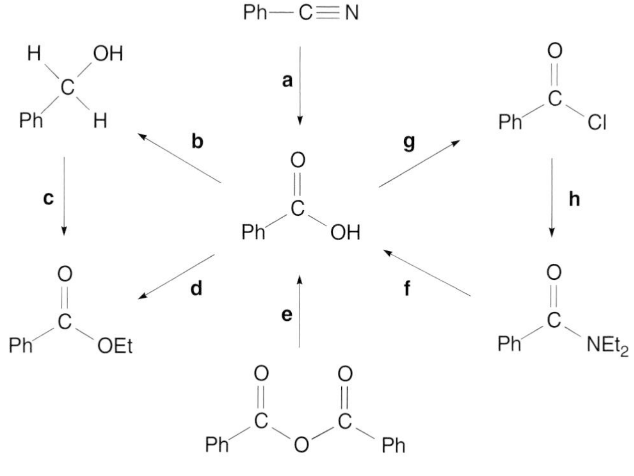

Section 9.11

3. Give the major condensation products from reaction of each of the following starting materials with sodium ethoxide (NaOEt) and then aqueous acid.
   (a) $PhCOCH_3 + PhCO_2CH_2CH_3$
   (b) $CH_3CH_2OCHO + PhCOCH_3$
   (c) $CH_3CO_2CH_2CH_3 + (CH_3)_3CCO_2CH_2CH_3$
   (d) $PhCOCH_3 + CH_3CO_2CH_2CH_3$

Sections 9.4.2, 9.7, 6.2.2.7 and 8.3.7.2

4. Heating (S)-2-hydroxybutyric acid with methanol and an acid catalyst gives compound **E**. On reaction of **E** with lithium aluminium hydride followed by water, compound **F** is formed which reacts with periodic acid (HIO_4) to give **G**. Finally, on reaction of **G** with 2,4-dinitrophenylhydrazine (Brady's reagent) and sulfuric acid, a precipitate of compound **H** ($C_9H_{10}N_4O_4$) is formed. Propose structures for compounds **E-H**.

5. The following questions are based on the reaction scheme below.

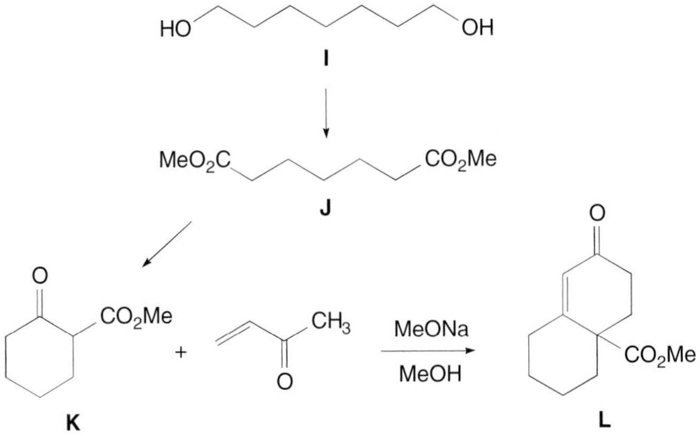

Section 8.3.3.5 and 9.4.2
Section 9.11.3
Sections 8.5.3 and 8.5.4

(a) Suggest reagents for converting **I** into **J** in two steps.
(b) Suggest reagents for converting **J** into **K**.
(c) Propose a mechanism for the conversion of **K** into **L**. (Hint: remember that enones can react with nucleophiles in Michael-type additions).

6. Draw structures for compounds **M**, **N** and **O**.

Sections 8.3.3.1, 8.5.4, 9.3.3, 9.5 and 9.11.3

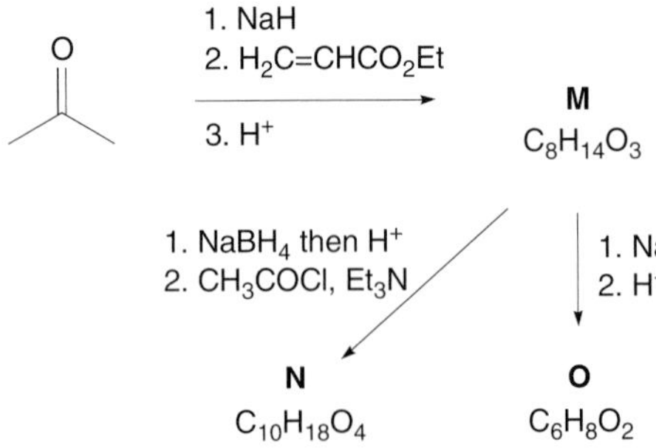

10

Spectroscopy

Key point. Spectroscopy is used to determine the structure of compounds. Spectroscopic techniques *include mass spectrometry* (MS), *ultraviolet spectroscopy* (UV), *infrared spectroscopy* (IR) and *nuclear magnetic resonance spectroscopy* (NMR). MS provides information on the size and formula of compounds by measuring the mass-to-charge ratio of organic ions produced on electron bombardment. UV, IR and NMR spectroscopy rely on the selective absorption of electromagnetic radiation by organic molecules. UV spectroscopy provides information on conjugated π-systems, while IR spectroscopy shows what functional groups are present. The most important method for structure determination is NMR, which can provide information on the arrangement of hydrogen and carbon atoms within an organic molecule.

^1H *NMR Chemical Shifts*

$$\underset{\underset{2.20\ ppm\quad 4.10\ ppm}{H_3C}}{\overset{\overset{O}{\|}}{C}}\underset{}{OCH_2CH_3}\qquad 1.30\ ppm$$

^{13}C *NMR Chemical Shifts*

$$170\ ppm\underset{\underset{20\ ppm\quad 60\ ppm}{H_3C}}{\overset{\overset{O}{\|}}{C}}\underset{}{OCH_2CH_3}\qquad 15\ ppm$$

10.1 Mass spectrometry (MS)

10.1.1 Introduction

A mass spectrometer converts organic molecules to positively charged ions, sorts them according to their mass-to-charge ratio (m/z), and determines the relative amounts of the ions present. In electron impact (EI) MS, a small sample is introduced into a high vacuum chamber where it is converted into a vapour and bombarded with high-energy electrons. These bombarding electrons eject an electron from the molecule (M) to give a radical cation ($M^{+\bullet}$), which is called the *parent peak* or *molecular ion*.

Note that a radical cation has a positive charge and an unpaired electron

$$R-H \xrightarrow[\text{bombardment}]{\text{electron}} \left[R-H\right]^{\oplus\bullet} + e^{\ominus}$$

molecule   radical cation

R is commonly used to represent an alkyl group (Section 2.2)

Keynotes in Organic Chemistry, Second Edition. Andrew F. Parsons.
© 2014 John Wiley & Sons, Ltd. Published 2014 by John Wiley & Sons, Ltd.

The electron that is ejected from the molecule will be of relatively high energy, for example, from a lone pair, which is not involved in bonding.

Examples

Functional groups, such as amines and ketones, are introduced in Section 2.1

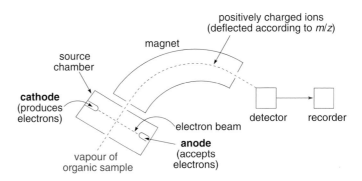

The molecular ions then pass between the poles of a powerful magnet, which deflects them (the deflection depends on the mass-to-charge ratio, m/z), before hitting an ion detector. Since the molecular ion has a mass that is essentially identical to the mass of the molecule, mass spectrometers can be used to determine the relative molecular mass.

The weakest bond is broken in a radical cation, so as to form the most stable fragment cation and fragment radical

If the bombarding electrons have enough energy, this can lead to *fragmentation* of the molecular ion into smaller radicals and cations (called daughter ions). Only charged particles (i.e. radical cations and cations) can be recorded by the detector.

A mass spectrum determines the masses of the radical cation and cations and their relative concentrations. The most intense peak is called the *base peak* and this is assigned a value of 100%. The intensities of the other peaks are reported as percentages of the base peak. The base peak can be the molecular ion peak or a fragment peak.

10.1.2 Isotope patterns

The molecular ion peak (M) is usually the highest mass number except for the isotope peaks, e.g. (M + 1), (M + 2), etc. Isotope peaks are present because many molecules contain heavier isotopes than the common isotopes (e.g. ^{13}C rather than ^{12}C).

Example: methane (CH_4)

$\left[CH_4 \right]^{\oplus \bullet}$

peak	M	M+1	M+2
formula	$^{12}C^1H_4$	$^{13}C^1H_4$	$^{13}C^2H^1H_3$
m/z	16	17	18
% of base peak	100	1.14	negligible

The isotope peaks (M + 1) and (M + 2) are much less intense than the molecular ion peak because of the natural abundance of the isotopes, e.g. $^{12}C = 98.9\%$ and $^{13}C = 1.1\%$. If only C, H, N, O, F, P, I are present, the *approximate* intensities of the (M + 1) and (M + 2) peaks can be calculated as follows.

^{15}N, ^2H and ^{17}O, as well as ^{13}C, can also contribute to the M + 1 peak

% (M + 1) ≈ 1.1 × number of C atoms + 0.36 × number of N atoms

$$\% \ (M + 2) \approx \frac{(1.1 \times \text{number of C atoms})^2}{200} + 0.20 \times \text{number of O atoms}$$

- The presence of one *chlorine* atom in a molecule can be recognised by the characteristic 3 : 1 ratio of (M) and (M + 2) peaks in the mass spectrum. This is due to ^{35}Cl (75.8%) and ^{37}Cl (24.2%) isotopes.
- The presence of one *bromine* atom in a molecule can be recognised by the characteristic 1 : 1 ratio of (M) and (M + 2) peaks in the mass spectrum. This is due to ^{79}Br (50.5%) and ^{81}Br (49.5%) isotopes.

Functional groups such as chloroalkanes (RCl) and acyl chlorides (RCOCl) contain chlorine (Section 2.1)

- The presence of *nitrogen* in a molecule can be deduced from the *nitrogen rule*. This states that a molecule of even-numbered relative molecular mass must contain either no nitrogen or an even number of nitrogen atoms, while a molecule of odd-numbered relative molecular mass must contain an odd number of nitrogen atoms.

Functional groups such as amines (e.g. RNH_2) and amides (RCONH$_2$) contain nitrogen (Section 2.1)

10.1.3 Determination of molecular formula

It is often possible to derive the molecular formula of a compound by recording a high-resolution mass spectrum. As the atomic masses of the isotopes have non-integral masses (except ^{12}C), the measurement of accurate masses to four to five decimal places allows compounds with different atomic compositions to be distinguished.

Example

Compound	$^{12}C^{16}O$	$^{14}N_2$	$\leftarrow$ *different*
Low resolution *m/z*	28	28	$\leftarrow$ *same*
High resolution *m/z*	$\begin{array}{l}^{12}C = 12.0000 \\ \underline{^{16}O = 15.9949} \\ 27.9949\end{array}$	$\begin{array}{l}^{14}N = 14.0031 \\ \underline{^{14}N = 14.0031} \\ 28.0062\end{array}$	$\leftarrow$ *different*

From the molecular formula, the number of *sites* (or *degrees*) *of unsaturation* of an unknown molecule can also be calculated. This number is equal to the sum of the number of rings, the number of double bonds and twice the number of triple bonds. The sites of unsaturation for compounds containing C, H, N, X (halogen), O and S can be calculated from the following formula.

$$\text{sites of unsaturation} = \text{carbons} - \frac{\text{hydrogens}}{2} - \frac{\text{halogens}}{2} + \frac{\text{nitrogens}}{2} + 1$$

Example

$$C_7H_7OCl \qquad \text{sites of unsaturation} = \frac{7 - 7 - 1 + 1}{2 \quad 2} = 4$$

This compound is called 1-chloro-4-methoxybenzene. Naming compounds is discussed in Section 2.4

For example: $Cl \!\!-\!\!\bigcirc\!\!-\!\! OCH_3$

10.1.4 Fragmentation patterns

On electron impact (EI), molecules first dissociate at the weaker bonds. The resulting fragments may undergo further fragmentation and analysis of the fragmentation pattern can provide information about the structure of the parent molecule. There is a series of general guidelines which can be used to predict prominent fragmentation pathways. These rely on the formation of the most stable cation.

Electron-donating (+I) alkyl groups (R) will stabilise a positive charge (Section 4.3)

1. Fragmentation of *alkanes* is most likely at highly substituted sites because of the increased stability of tertiary (R_3C^+) carbocations over secondary (R_2CH^+) and particularly primary (RCH_2^+) carbocations (see Section 4.3).

Example

cleavage here
generates the most
stable carbocation

*tertiary
carbocation*

$^+C(CH_3)_3$ is called the *tert*-butyl cation

$CH_3CH_2CH_2^{\bullet}$ is called the 1-propyl radical

2. Fragmentation of *halogenoalkanes* (R–X) usually leads to cleavage of the weak carbon–halogen bond to form a carbocation (R^+) and a halogen radical (or atom, $X^{\bullet}$).

3. Fragmentation of *alcohols* (ROH), *ethers* (ROR) or *amines* (e.g. RNH_2) usually leads to cleavage of the C–C bond next to the heteroatom to generate a resonance-stabilised carbocation. This is called α-cleavage.

In S_N1 reactions, R–X reacts to form R^+ and X^- (Section 5.3.1.2)

Resonance stabilisation of carbocations is discussed in Section 1.6.3

Example

Molecular ion of diethyl ether

α-cleavage

Notice the use of single-headed curly arrows (Section 4.1)

4. Fragmentation of *carbonyl compounds* usually leads to cleavage of the C(O)–C bond to generate a resonance-stabilised carbocation. This is called α-cleavage.

Example

Molecular ion of butanone

α-cleavage

The RCO^+ ion is called an acylium ion (see Section 7.2.5)

10.1.5 Chemical ionisation (CI)

Extensive fragmentation can occur in EI (electron-impact) spectra because the bombarding electron imparts very high energy to the molecular ion, and fragmentation provides a way to release the energy. A less energetic ionisation technique, which can result in less fragmentation of the molecular ion, is *chemical ionisation* (CI). In this approach, the organic sample is protonated in the gas phase to give a cation (not a radical cation as for EI) and the $[M + H]^+$ peak is detected by the mass spectrometer (i.e. one mass unit higher than the molecular mass).

10.2 The electromagnetic spectrum

Ultraviolet (UV), infrared (IR) and nuclear magnetic resonance (NMR) spectroscopy involve the interaction of molecules with electromagnetic radiation. When an organic molecule is exposed to electromagnetic energy of different wavelengths, and hence different energies (see the equation below), some of the wavelengths will be absorbed, and this can be recorded in an *absorption spectrum*. It should be noted that high frequencies, large wavenumbers and short wavelengths are associated with high energy.

$$E = h\nu = \frac{hc}{\lambda} = h\bar{\nu}c$$

E = energy (J mol^{-1})
h = Planck's constant (J s)
ν = frequency (s^{-1} or Hz)
λ = wavelength (m)
c = velocity of light (m s^{-1})
$\bar{\nu}$ = wavenumber (cm^{-1})

The absorption of radiation depends on the structure of the organic compound and the wavelength of the radiation. Different wavelengths of radiation affect organic compounds in different ways. The absorption of radiation increases the energy of the organic molecule. This can lead to: (i) excitation of electrons from one molecular orbital to another in UV spectroscopy; (ii) increased molecular motions (e.g. vibrations) in IR spectroscopy; or (iii) excitation of a nucleus from a low nuclear spin state to a high spin state in NMR spectroscopy.

Type of spectroscopy	Radiation source	Energy range (kJ mol^{-1})	Type of transition
Ultraviolet	ultraviolet light	595–298	Electron excitation
Infrared	infrared light	8–50	Molecular vibrations
NMR	radio-waves	25–251 $\times$ 10^{-6}	Nuclear spin

10.3 Ultraviolet (UV) spectroscopy

Orbitals are discussed in Section 1.4

The organic molecule is irradiated with ultraviolet radiation of changing wavelength and the absorption of energy is recorded. The absorptions correspond to the energy required to excite an electron to a higher energy level (e.g. from an occupied orbital to an unoccupied or partially occupied orbital).

Ultraviolet (UV) spectrometer

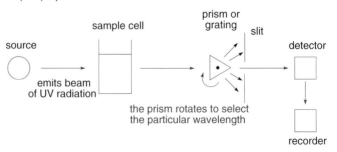

The exact amount of UV radiation absorbed at a particular wavelength is expressed as the compound's *molar absorptivity* or *molar absorption coefficient* (ε). This is a measure of the efficiency of radiation absorption and is calculated from the absorbance of radiation, which is derived from the *Beer-Lambert Law*.

$$\varepsilon = \frac{A}{c \times l}$$ ε = molar absorptivity ($m^2\ mol^{-1}$) A = absorbance c = sample concentration (mol dm^{-3}) l = sample path length (cm)	The Beer-Lambert Law: $A = log_{10}\left(\dfrac{I_0}{I}\right)$ I_0 = intensity of incident radiation striking sample I = intensity of transmitted radiation emerging from sample

The intensity of an absorption band is usually quoted as the molar absorption coefficient at maximum absorption, ε_{max}. On absorption of UV radiation, electrons in bonding or nonbonding orbitals (ground state) can be given sufficient energy to transfer to higher energy antibonding orbitals (excited state). The most common excitation of electrons involve $\pi \rightarrow \pi^*$ and $n \rightarrow \pi^*$ transitions.

π and π^* orbitals are discussed in Section 1.4

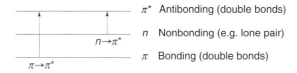

π^* Antibonding (double bonds)

n Nonbonding (e.g. lone pair)

π Bonding (double bonds)

The exact wavelength of radiation required to effect the $\pi \rightarrow \pi^*$ or $n \rightarrow \pi^*$ transition depends on the energy gap, which in turn depends on the nature of the functional group(s) present in the organic molecule. The absorbing group within a molecule, which is usually called the *chromophore*, can therefore be assigned from the wavelength of the absorption peak (λ_{max}).

Functional groups are discussed in Section 2.1

Chromophore	Transition	λ_{max} (nm)
$-\overset{..}{\underset{..}{O}}-$	$n \rightarrow \pi^*$	≈ 185
$C=C$	$\pi \rightarrow \pi^*$	≈ 190
$C=\overset{..}{\underset{..}{O}}$	$n \rightarrow \pi^*$	≈ 300

One of the most important factors that affects λ_{max} is the extent of conjugation (e.g. the number of alternating C–C and C=C bonds). The greater the number of conjugated double bonds, the larger the value of λ_{max}. If a compound has enough double bonds, it will absorb visible light ($\lambda_{max} > 400$ nm) and the compound will be coloured. The UV spectrum can therefore provide information on the nature of any conjugated π system.

Conjugation is introduced in Section 1.6.3

Examples

Naming carbon chains is introduced in Section 2.4; for assigning *E,Z* nomenclature, see Section 3.3.1.2

2-propenal	1,3-butadiene	(*E*)-hexa-1,3,5-triene
λ_{max} = 210 nm ($\pi \to \pi^*$)	λ_{max} = 217 nm	λ_{max} = 258 nm

An *auxochrome* is a substituent such as NH_2 and OH which, when attached to a chromophore, alters the λ_{max} and the intensity of the absorption. For example, whereas benzene (C_6H_6) has a λ_{max} at 255 nm, phenol (C_6H_5OH) has a λ_{max} at 270 nm and aniline ($C_6H_5NH_2$) has a λ_{max} at 280 nm. There is a shift to longer wavelengths because the lone pair of electrons on oxygen and nitrogen can interact with the π system of the benzene ring. Whereas a shift to longer wavelengths is called a *red shift*, a shift to lower wavelengths is called a *blue shift*.

The lone pair on the nitrogen atom of aniline is stabilised by delocalisation (see Section 1.7.2)

10.4 Infrared (IR) spectroscopy

The organic molecule (solid, gas or liquid) is irradiated continuously with infrared radiation (of changing wavelength) and the absorption of energy is recorded by an IR spectrometer (which has the same design as a UV spectrometer, Section 10.3). The absorption corresponds to the energy required to vibrate bonds within a molecule.

The absorption of energy, which gives rise to bands in the IR spectrum, are reported as frequencies and these are expressed in *wavenumbers* (in cm^{-1}). The most useful region of radiation is between 4000–400 cm^{-1}.

The electromagnetic spectrum is discussed in Section 10.2

$$\text{wavenumber } (\bar{v}) = \frac{1}{\text{wavelength } (\lambda)}$$

The frequency of vibration between two atoms depends on the strength of the bond between them and on their atomic weights (*Hooke's Law*). A bond can only stretch, bend or vibrate at specific frequencies corresponding to specific energy levels. If the frequencies of the IR radiation and the bond vibration are the same, then the vibrating bond will absorb energy.

$$\bar{v} = \frac{1}{2\pi c} \sqrt{k\frac{(m_1 + m_2)}{m_1 m_2}}$$

$\bar{v}$ = vibrational wavenumber (cm^{-1})
k = force constant, indicating the bond strength (N m^{-1})
$m_1 m_2$ = masses of atoms (kg)
c = velocity of light (cm s^{-1})

The IR spectrum of an organic molecule is complex because all the bonds can stretch and also undergo bending motions. Those vibrations that lead to a change in dipole moment are observed in the IR spectrum. Bending vibrations generally occur at lower frequencies than stretching vibrations of the same group.

Dipole moments are discussed in Section 1.6.1

The IR spectrum can therefore be viewed as a unique fingerprint of an organic compound and the region below $1500\,\mathrm{cm}^{-1}$ is called the *fingerprint region*. Fortunately, the vibrational bands of functional groups in different compounds do not change much, and they appear at characteristic wavenumbers. These bands, particularly stretching vibrations above $1500\,\mathrm{cm}^{-1}$, can provide important structural information.

Functional groups are introduced in Section 2.1

Bond or functional group	$\bar{v}$ (cm^{-1})
RO–H, C–H, N–H	4000–2500
RC≡N, RC≡CR	2500–2000
C=O, C=N, C=C	2000–1500
C–C, C–O, C–N, C–X	<1500

In general, short strong bonds vibrate at higher frequency than long weak bonds (as more energy is required). Therefore the C≡C bond absorbs at a higher wavenumber than C=C, which, in turn, absorbs at a higher wavenumber than C–C.

In general, bonds bearing light atoms vibrate at higher frequency than bonds bearing heavier atoms. Therefore the C–H bond absorbs at a higher wavenumber than the C–C bond.

Alcohols and amines

The intense O–H or N–H stretching vibration between 3650 and $3200\,\mathrm{cm}^{-1}$ is the most characteristic peak of alcohols (ROH) or amines (RNH$_2$ or R$_2$NH), respectively. If this peak is broad, then this often indicates intermolecular hydrogen bonding.

Examples

Naming carbon chains is introduced in Section 2.4

~3400 cm^{-1}
NH$_2$

~1650 cm^{-1} ~3300 cm^{-1}
OH

~1600 cm^{-1}

2-propenol   aniline

Carbonyl compounds

The carbonyl functional group exhibits a sharp, intense peak between 1575 and 1825 cm^{-1} due to the C=O stretching vibration. The exact position of the peak can be used to identify the type of carbonyl compound.

Functional groups are introduced in Section 2.1

Carbonyl	$\bar{v}$ / cm^{-1}	Carbonyl	$\bar{v}$ / cm^{-1}
acid anhydride (RCO$_2$COR)	1820	ketone (RCOR)	1715
acyl chloride (RCOCl)	1800	carboxylic acid (RCO$_2$H)	1710
ester (RCO$_2$R)	1740	amide (e.g. RCONH$_2$)	1650
aldehyde (RCHO)	1730		

Examples

The electronic properties of the groups attached to the C=O bond affect the stretching vibration

~1715 cm^{-1} ~1730 cm^{-1} ~1740 cm^{-1} ~1800 cm^{-1}
O   O   O   O

butanone   ethanal   ethyl ethanoate   ethanoyl chloride

10.5 Nuclear magnetic resonance (NMR) spectroscopy

When certain nuclei of an organic molecule (e.g. ^{1}H, ^{13}C, ^{19}F and ^{31}P) are placed within a strong magnetic field, the nuclear spins align themselves with (parallel to) or against (antiparallel to) the field. Those that are parallel to the field are slightly favoured because they are lower in energy. On irradiation of electromagnetic radiation of the correct frequency, energy can be absorbed to produce *resonance*. This results in 'spin flipping' and the lower energy nuclei are promoted to the higher energy state. The nuclei can then relax to their original state by releasing energy.

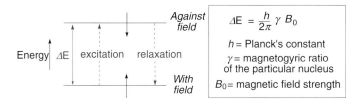

The energy difference (ΔE) between the two states depends on the magnetic field strength (B_0) and the type of nucleus (γ). These in turn determine the exact amount of radio-frequency energy (v) which is required for resonance (as $\Delta E = hv$). Thus, the smaller the magnetic field, the smaller the energy difference, which means that lower frequency (and lower energy) radiation is needed. The absorption frequency (v) for ^{1}H and ^{13}C nuclei within the same molecule, are not all the same.

Electrons surround the nuclei and produce small local induced magnetic fields (B_i) that act against the applied magnetic field (B_0). The magnetic field actually felt by the nucleus is therefore a little smaller than the applied field and the nuclei are said to be *shielded*. Therefore, the more electron density there is near a nucleus, the greater will be the shielding.

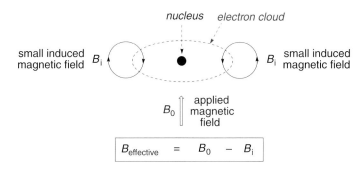

- *Electronegative* groups near a nucleus will pull electrons away and *decrease* the shielding of the nucleus.
- *Electropositive* groups near a nucleus will increase the surrounding electron density and *increase* the shielding of the nucleus.

Electronegative atoms, or groups, attract the electron density; electropositive atoms (or groups) repel the electron density (Section 1.6.1)

As each type of nucleus has a slightly different electronic environment, each nucleus will be shielded to a slightly different extent. A high-resolution NMR spectrometer can detect the small differences in the effective magnetic fields of the nuclei and produce different NMR resonance signals, or peaks, for each type of nuclei.

NMR spectrometer

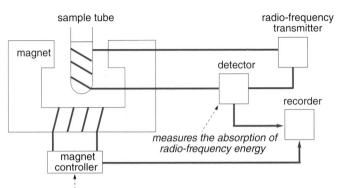

The NMR spectrum therefore records the difference in effective field strength (horizontal axis) against the intensity of absorption of radiofrequency energy (vertical axis). In older machines, this is achieved by keeping the radiofrequency constant and varying the strength of the applied magnetic field. The high-field (or upfield) side is on the right while the low-field (or downfield) side is on the left of the spectrum.

Peaks on the left have a high chemical shift, those on the right have a low chemical shift

The intensity of the peak is usually not shown on the spectrum

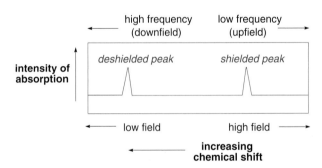

In modern instruments, the magnetic field is kept constant and the radiofrequency is varied in *pulse-Fourier transform NMR* (FT-NMR). In FT-NMR all of the nuclear spins are excited instantaneously using a mixture of radiofrequencies. The spectrum is obtained by analysing the emission of radiofrequency energy (as the spins return to equilibrium) as a function of time.

The NMR spectrum is calibrated using tetramethylsilane (Me$_4$Si), which produces a single low frequency peak in both ^{1}H and ^{13}C NMR spectra. This is set as a *chemical shift* value of 0 on the delta (δ) scale, where $1 \delta = 1$ part per million (ppm) of the operating frequency of the spectrometer. Almost all other absorptions occur at higher frequency to this signal, typically 0–10 ppm for ^{1}H NMR and 0–210 ppm for ^{13}C NMR.

$$\delta \, (\text{ppm}) = \frac{\text{distance of peak from Me}_4\text{Si (Hz)} \times 10^6}{\text{spectrometer frequency (Hz)}}$$

^{1}H and ^{13}C nuclei are the most commonly observed nuclei but because they absorb in different radiofrequency regions, they cannot be observed at the *same* time (i.e. in the ^{1}H NMR spectrum, ^{13}C signals are not observed).

10.5.1 ^{1}H NMR spectroscopy

To obtain a ^{1}H NMR (or proton NMR) spectrum, a small amount of the sample is usually dissolved in a deuterated solvent (e.g. $CDCl_3$) and this is placed within a powerful magnetic field. The spectrum can provide information on the number of equivalent hydrogens in an organic molecule. Equivalent hydrogens show a single peak while non-equivalent hydrogens give rise to separate peaks. The number of peaks in the spectrum can therefore be used to determine how many different kinds of hydrogen are present. The relative number of hydrogen atoms responsible for the peaks in the ^{1}H NMR spectrum can be determined by *integration* of the peak areas. The areas of the peaks are recorded on the spectrum as integration curves.

Example

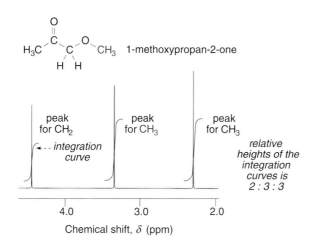

In 1-methoxypropan-2-one there are three different types of hydrogens, each producing a peak in the ^{1}H NMR spectrum

10.5.1.1 Chemical shifts

The chemical shift (δ) value of the peak provides information on the magnetic/chemical environment of the hydrogens. Hydrogens next to electron-withdrawing groups are *deshielded* (leading to high δ values) whereas hydrogens next to electron-donating groups are *shielded* (leading to low δ values).

The structure of halogenoalkanes (RX) is discussed in Section 5.1

Example

~4.5 ppm   ~3.5 ppm   ~3.4 ppm   ~3.2 ppm

R–CH$_2$–F   R–CH$_2$–Cl   R–CH$_2$–Br   R–CH$_2$–I

Decreasing electronegativity of the halogen atom ⟶

Functional groups therefore have characteristic chemical shift values.

Hydrogen atoms close to an electron-withdrawing atom or group have relatively high chemical shifts

For tables of ^1H NMR chemical shift values, see Appendix 6

Functional group	δ (ppm)
alkane CH–C	0–1.5
allylic, benzylic CH–C=C	1.5–2.5
halogenoalkane CH–X amine CH–NR_2 ether CH–OR alcohol CH–OH	2.5–4.5
alkene CH=C	4.5–6.5
aromatic CH=C	6.5–8
aldehyde CH=O	9–10

Example

Hydrogen atoms close to two functional groups (e.g. a benzene ring and bromine atom) are affected by both groups

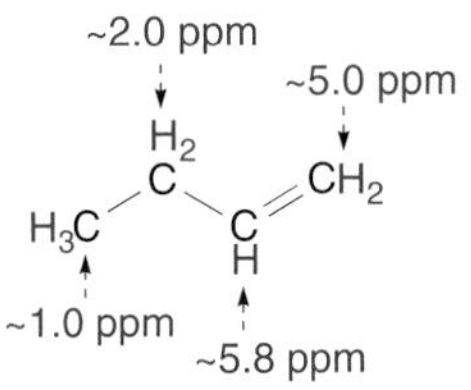

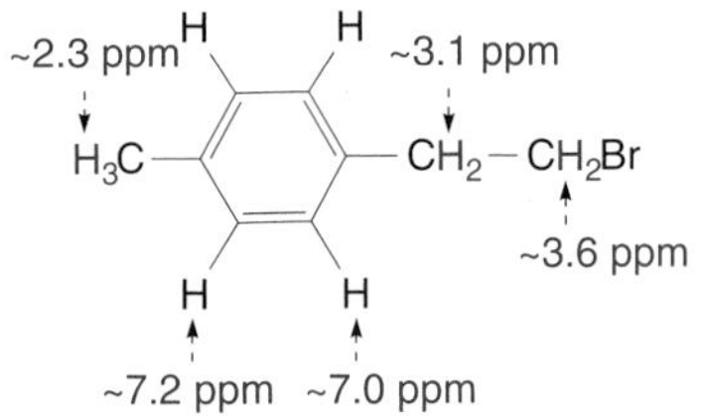

In general, the peaks for methyl (CH_3) hydrogens appear at low chemical shift values, while methylene (CH_2) hydrogens appear at slightly higher values and methine hydrogens (CH) at even higher values.

Example

In each case, the hydrogens closest to the electron-withdrawing oxygen have the higher chemical shifts

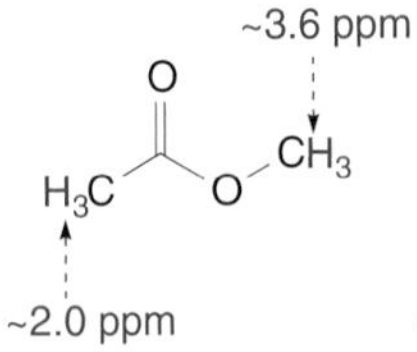

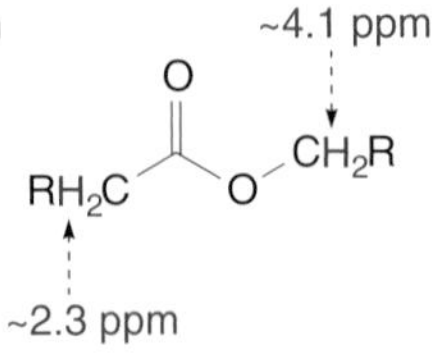

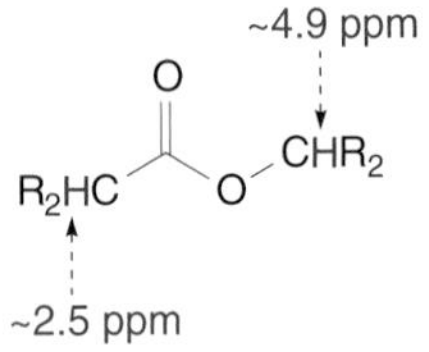

Aromatics, alkenes and aldehydes

Hydrogens bonded to an aromatic ring are strongly deshielded and absorb downfield. When the π electrons enter the magnetic field, they circulate around the ring to generate a *ring current*. This produces a small, induced magnetic field that reinforces the applied field outside the ring, resulting in the aromatic hydrogens being deshielded. The presence of an aromatic ring current is characteristic of aromatic compounds. A related effect is observed for alkenes (RCH=CHR) and aldehydes (RCHO). For these compounds, circulation of π-electrons in the double bonds produces induced magnetic fields. These are responsible for the high chemical shift values of alkene hydrogens and, particularly, aldehyde hydrogens.

The structure of benzene and other aromatic compounds is discussed in Section 7.1

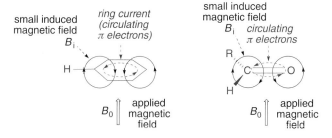

10.5.1.2 Spin-spin splitting or coupling

Peaks in the ^{1}H NMR spectrum are often not singlets (i.e. single peaks) because of *spin-spin splitting or coupling*. The small magnetic field of one nucleus affects the magnetic field of a neighbouring nucleus and this 'coupling' results in splitting of the peaks. The distance between the peaks is called the coupling constant (J), which is measured in Hertz (Hz).

The appearance of the peak depends on the number of neighbouring hydrogens. This can be calculated using the $n+1$ rule, where n is the number of equivalent neighbouring hydrogens. The multiplicity and relative intensities of the peaks can be obtained from Pascal's triangle (shown below).

Typically, hydrogens on the same carbon are equivalent and so do not split each other

n	number of peaks ($n+1$)	peak pattern	integration ratios
0	1	singlet (s)	1
1	2	doublet (d)	1 : 1
2	3	triplet (t)	1 : 2 : 1
3	4	quartet (q)	1 : 3 : 3 : 1
4	5	quintet (quin)	1 : 4 : 6 : 4 : 1
5	6	sextet (sex)	1 : 5 : 10 : 10 : 5 : 1
6	7	septet (sep)	1 : 6 : 15 : 20 : 15 : 6 : 1

For CH_3CH_2-, the CH_3 peak appears as a triplet (1:2:1) and the CH_2 peak as a quartet (1:3:3:1)

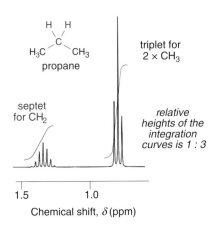

The two methyl groups in propane
are chemically equivalent

For spin-spin splitting to occur, the neighbouring hydrogen atoms must be chemically (and magnetically) non-equivalent. One simple way of determining whether hydrogens are equivalent or non-equivalent involves mentally substituting the hydrogens with a phantom group Z. For example, replacement of either of the hydrogen atoms in CH_2Cl_2 with Z will form the same molecule ($ZCHCl_2$) and so the hydrogen atoms are equivalent (see below). For $Me(Cl)C=CH_2$, however, replacement of either of the alkene hydrogen atoms with Z will form alkene diastereoisomers, which are not identical. The two hydrogen atoms are therefore not equivalent and will give different signals in the NMR spectrum.

Diastereoisomers are introduced in
Section 3.3.2.4

The use of solid and dashed lines is
discussed in Section 3.3.2

identical (H = H)

Replace H with Z

Replace H with Z

not identical (H ≠ H)

Replace H with Z

Replace H with Z

Spin-spin splitting can be observed between non-equivalent hydrogen atoms that are on the same carbon atom (geminal coupling, $J \approx 10\text{–}20\,\text{Hz}$) or adjacent carbon atoms (vicinal coupling, $J \approx 7\,\text{Hz}$). Coupling is not usually observed between hydrogens separated by more than three σ bonds. It should be noted that hydrogens which are coupled to each other must have the *same J* value.

Sigma bonds are introduced in
Section 1.4

Alkenes

cis- and trans-Isomerism is
discussed in Section 3.3.1.1

The size of the coupling constant (J) depends on the dihedral angle (ϕ) between hydrogens on adjacent carbon atoms (this is called the Karplus relationship). This can be used to identify *cis*- and *trans*-isomers of disubstituted alkenes, $RCH=CHR$.

- For *cis*-alkenes, the dihedral angle between the two C–H bonds is 0° (as the two hydrogen atoms are on the same side of the double bond) and $J = 7$–11 Hz.
- For *trans*-alkenes, the dihedral angle between the two C–H bonds is 180° (as the two hydrogen atoms are on the opposite side of the double bond) and $J = 12$–18 Hz.

Dihedral angles are discussed in Section 3.2.1

Aromatics

For benzene (C_6H_6), the splitting between two *ortho* (1,2-) hydrogen atoms is ca. 8 Hz. For *meta* (1,3-) and *para* (1,4-) hydrogen atoms the splitting is much lower ($J_{meta} = 1$–3 Hz, $J_{para} = 0$–1 Hz) because the atoms are further apart and these are known as *long-range couplings*. Different types of hydrogens often produce signals that overlap to give multiplets (m). This is particularly true for hydrogens of mono-substituted benzenes (C_6H_5–R), which are generally observed as a single overlapping (broad) absorption.

The terms *ortho*-, *meta*- and *para*- are defined in Section 2.4

Alcohols

The OH hydrogen of alcohols (ROH) is generally observed as a single (sharp or broad) peak even when a vicinal proton (C<u>H</u>OH) is present. This is because the alcohol OH hydrogens undergo fast proton exchange with each other (or with traces of water in the sample). This averages the local field leading to *decoupling*. Indeed, the peak due to the OH group of the alcohol can be removed in the ^{1}H NMR spectrum by simply shaking a solution of the alcohol with D_2O (to form ROD and HOD). The fact that the peak for ROH disappears and a new peak at ~4.7 ppm arises for HOD can be used to show the presence of an OH (or NH) group.

A deuterium atom is not detected in a ^{1}H NMR spectrum

10.5.1.3 Summary

The ^{1}H NMR spectrum can provide the following structural information:

1. the types of chemically different hydrogens (from the number of peaks and the chemical shift values);
2. the number of each type of hydrogen (from integration of the peak areas);
3. the number of nearest (hydrogen) neighbours that each hydrogen has (from the appearance of the peak, which is determined by spin-spin splitting).

Example

hydrogen	δ / ppm	peak appearance
H_A	7.64*	doublet ($J = 16$ Hz)
H_B	7.30–7.14	broad multiplet
H_C	6.39	doublet ($J = 16$ Hz)
H_D	4.19	quartet ($J = 7$ Hz)
H_E	1.30	triplet ($J = 7$ Hz)

ethyl (*E*)-cinnamate

*This alkene hydrogen is deshielded because of conjugation with the (electron-withdrawing) carbonyl group.

Assignment of *E* and *Z* configuration is discussed in Section 3.3.1.2

Conjugation is discussed in Section 1.6.3

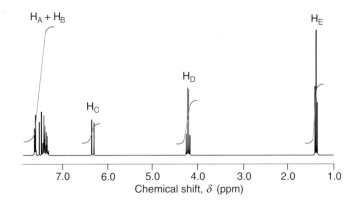

Notice that the doublet for H_C leans towards the doublet for H_A– this is called the roof effect

10.5.2 ^{13}C NMR spectroscopy

The ^{13}C NMR spectrum can provide information on the number of different types of carbon atoms in an organic molecule. Carbons, which are equivalent, show a single absorption (or peak), while non-equivalent carbons give rise to separate absorptions.

As ^{13}C has a natural abundance of only 1.1%, the signals are around 1/6000 as strong as those for ^{1}H. However, spectra can be routinely recorded by Fourier transform (FT) NMR using short pulses of radiofrequency radiation. *Broad-band proton decoupling* is usually employed which removes all ^{13}C–^{1}H coupling so that singlet peaks are observed for each different kind of carbon. (As most ^{13}C are surrounded by ^{12}C, ^{13}C–^{13}C coupling is minimal.)

The chemical shift (δ) values of the peaks provide information on the magnetic/chemical environment of the carbons. As for ^{1}H NMR, carbons next to electron-withdrawing groups produce high frequency signals (leading to high δ values), whereas carbons next to electron-donating groups produce low frequency signals (leading to low δ values).

Hybridisation is discussed in Section 1.5

Functional groups have characteristic carbon chemical shifts and these generally follow the same trends as for hydrogen chemical shifts. Hence, the order of chemical shifts is usually tertiary (CH) > secondary (CH$_2$) > primary (CH$_3$) and sp^3 carbon atoms generally absorb between 0 and 90 ppm, while sp^2 carbon atoms absorb between 100 and 210 ppm. The chemical shift range is therefore much larger than for ^{1}H NMR, and so ^{13}C peaks are less likely to overlap.

Functional groups are discussed in Section 2.1

Alkane carbons	δ(ppm)
CH$_3$	5–20
CH$_2$	20–30
CH	30–50
C	30–45

Functional group	δ(ppm)
halogenoalkane C–X	0–80
alcohol/ether C–O	40–80
alkene C=	100–150
aromatic C=	110–160
carbonyl C=O	160–210

Unless special techniques are used, the area under a ^{13}C signal is *not* proportional to the number of carbon atoms giving rise to the signal.

Carbonyls

The carbon atom of the carbonyl group (C=O) is strongly deshielded and has a characteristic chemical shift of around 160–210 ppm. This chemical shift is higher than for any other type of carbon atom.

The *DEPT ^{13}C NMR spectrum* (distortionless enhanced polarisation transfer) identifies whether the carbon atom is primary (CH$_3$), secondary (CH$_2$), tertiary (CH) or quaternary (C).

The ^{13}C NMR spectrum can therefore provide the following structural information:

1. the types of chemically different carbons (from the number of peaks and the chemical shift values);
2. the number of hydrogens on each carbon (when using the DEPT technique).

Example

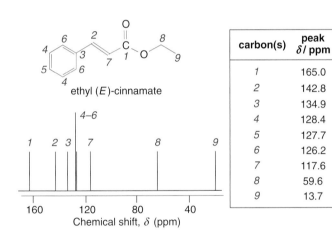

ethyl (*E*)-cinnamate

carbon(s)	peak δ/ ppm
1	165.0
2	142.8
3	134.9
4	128.4
5	127.7
6	126.2
7	117.6
8	59.6
9	13.7

The two carbon atoms at position 4, and the two at position 6, are chemically equivalent

Worked example

A strong smelling compound that is used as 'essence' of rum in food flavourings, was isolated and analysed using different spectroscopic techniques. The IR spectrum shows an intense absorption band at 1740 cm^{-1}, and the mass spectrum shows a base peak at *m/z* 57, due to a fragment cation. Use this information, together with the ^{1}H NMR spectrum shown below, to propose a structure for this compound.

Hint: Use the IR information to identify the functional group

Hint: Identify the structure of the fragment cation in the mass spectrum

Hint: From the ^{1}H NMR spectrum, use the chemical shift values and peak areas to identify partial structures. Use the splitting patterns to identify how the partial structures are joined together

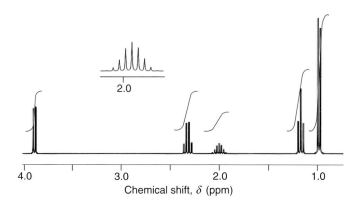

2.0

4.0   3.0   2.0   1.0

Chemical shift, δ (ppm)

Answer

The IR absorption band at $1740\,\text{cm}^{-1}$ indicates an ester (RCO_2R).

For IR spectroscopy see Section 10.4

In the mass spectrometer, an ester molecular ion fragments to cleave the C–O bond, to form an acylium ion RCO^+. The RCO^+ ion has a relative molecular mass of 57, so R has a relative molecular mass of 29 ($57 - 28 = 29$).

For fragmentation patterns in mass spectrometry, see Section 10.1.4

From the ^{1}H NMR spectrum the following partial structures are assigned to the five peaks:

3.86 ppm 2H possibly CH_2–OCOR
2.29 ppm 2H possibly CH_2–CO_2R
1.97 ppm 1H CH
1.14 ppm 3H CH_3
0.95 ppm 6H $2 \times CH_3$

For typical ^{1}H NMR chemical shift values for CH_3, CH_2 and CH groups, see Section 10.5.1.1 and Appendix 6

From the splitting patterns of each signal, the following partial structures can be joined together as follows:

3.86 ppm d –CH–CH_2–
2.29 ppm q –CH_2–CH_3
1.97 ppm 9 peaks –CH_2–$CH(CH_3)_2$
1.14 ppm t –CH_2–CH_3
0.95 ppm d –$CH(CH_3)_2$

For spin-spin splitting of ^{1}H NMR peaks, see Section 10.5.1.2

Connecting the partial structures together gives 2-methylpropyl propanoate. This compound is consistent with the MS data (the acylium ion is $CH_3CH_2CO^+$):

For naming esters, see Section 2.4

Problems

1. Compound **A** is believed to have the following structure.

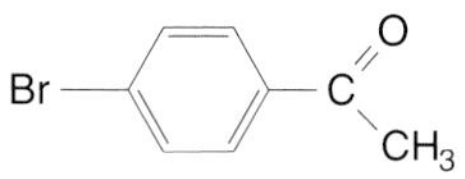

A

Sections 10.1 and 10.5

Explain how you would use NMR spectroscopy and/or mass spectrometry to:

(a) confirm the molecular formula;
(b) demonstrate the presence of a single bromine atom;
(c) establish the presence of the COCH$_3$ group;
(d) confirm the presence of a 1,4-disubstituted ring.

2. Propose a structure for compound **B** using the following spectroscopic information. Sections 10.1 and 10.5

m/z (EI) 134 (25%), 91 (85%), 43 (100%). (The percentages refer to peak heights.)

δ_H 7.40–7.05 (5H, broad multiplet), 3.50 (2H, s), 2.00 (3H, s). (CDCl$_3$ was used as the solvent.)

δ_C 206.0 (C), 134.3 (C), 129.3 (2 × CH), 128.6 (2 × CH), 126.9 (CH), 48.8 (CH$_2$), 24.4 (CH$_3$). (CDCl$_3$ was used as the solvent.)

3. Draw the ^{1}H and ^{13}C NMR (fully proton decoupled) spectra for compounds **C** and **D**, indicating approximate chemical shifts of the hydrogens and carbons and showing spin-spin splitting patterns in the ^{1}H NMR spectra. Sections 10.5.1 and 10.5.2

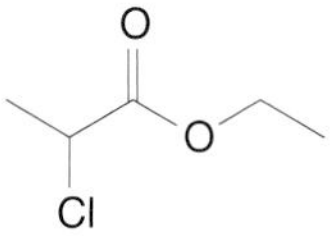

C 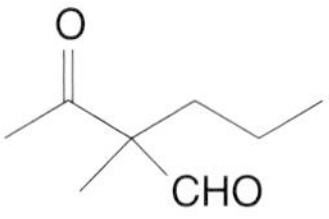   **D**

4. Compound **E** has a prominent IR absorption band at 1730 cm^{-1} and gives two singlet peaks in the ^{1}H NMR spectrum. On reaction with sodium borohydride followed by aqueous acid, compound **F** is obtained. Compound **F**, which has a broad IR absorption band at 3500–3200 cm^{-1}, reacts with sodium hydride (a base) followed by iodomethane to give compound **G**. The mass spectrum of **G** shows a molecular ion peak at m/z 102 and an intense peak at m/z 45. Propose structures for compounds **E–G**. Sections 10.1, 10.4 and 10.5.1

5. A compound with the molecular formula, C$_5$H$_{10}$O was isolated from carrot. Using the ^{1}H and ^{13}C NMR spectra provided below, propose a structure for this compound. Sections 10.5.1 and 10.5.2

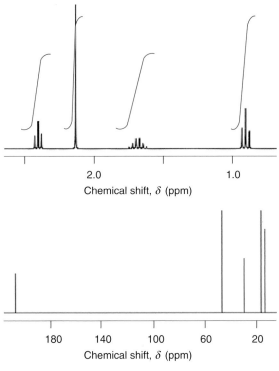

Sections 10.5.1 and 10.5.2

6. The ^{1}H and ^{13}C NMR spectra of a compound with the molecular formula $C_9H_{12}O$ are shown below. Use the spectra to propose a structure for this compound.

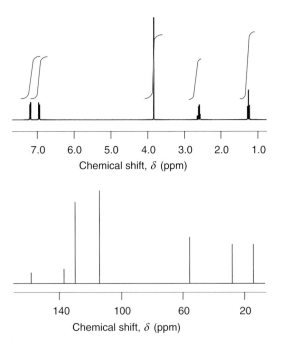

11

Natural products and synthetic polymers

Key point. Natural products are divided into particular classes of compound. These include *carbohydrates, lipids, amino acids, peptides and proteins,* and *nucleic acids.* Carbohydrates (or sugars) are poly-hydroxycarbonyl compounds that exist as monomers, dimers or polymers. Organic-soluble waxes, oils, fats and steroids are known as lipids, while the condensation of amino acids produces natural polyamides called peptides and proteins. Nucleic acids (which include RNA and DNA) are natural polymers made up of sugars, heterocyclic nitrogen bases and phosphate groups. Synthetic (unnatural) polymers, which have had a tremendous impact on our day-to-day living, can be classified as chain-growth (e.g. polyethylene) or step-growth polymers (e.g. polyamides or nylons).

D-glucose
a natural sugar

L-proline
a natural amino acid

polystyrene
a synthetic polymer

11.1 Carbohydrates

Carbohydrates are a class of naturally occurring polyhydroxylated aldehydes and ketones, which are commonly called sugars. Many, but not all, sugars have the empirical formula $C_X(H_2O)_y$.

- *Simple sugars,* or *monosaccharides,* are carbohydrates (such as glucose) which cannot be hydrolysed into smaller molecules.
- *Complex sugars* are carbohydrates, which are composed of two or more sugars joined together by oxygen bridges. These can be hydrolysed into their component sugars (e.g. sucrose is a disaccharide which is hydrolysed to one

Aldehydes and ketones are introduced in Section 2.1

A hydrolysis reaction is a reaction involving water

Keynotes in Organic Chemistry, Second Edition. Andrew F. Parsons.
© 2014 John Wiley & Sons, Ltd. Published 2014 by John Wiley & Sons, Ltd.

glucose molecule and one fructose molecule, while cellulose is a poly-saccharide which is hydrolysed to give around 3000 glucose molecules).

Monosaccharides bearing an aldehyde (RCHO) are called *aldoses*, whereas those bearing a ketone (RCOR) are called *ketoses*. Sugars bearing three, four, five and six carbon atoms are called trioses, tetroses, pentoses and hexoses, respectively.

Fischer projections are discussed in Section 3.3.2.6

CHO
H–C–OH
HO–C–H   ≡
H–C–OH
H–C–OH
CH₂OH

D-glucose (an aldohexose)

CHO
H——OH
HO——H
H——OH
H——OH
CH₂OH

Fischer projection

CH₂OH
═O
HO——H
H——OH
H——OH
CH₂OH

D-fructose
(a ketohexose)

Almost all carbohydrates are chiral and optically active. They are most often drawn as Fischer projections – the horizontal lines come out of the page while the vertical lines go into the page. If the chiral centre furthest from the aldehyde or ketone is the same absolute configuration as D-(+)-glyceraldehyde, then these are called D-sugars. (Almost all naturally occurring sugars are D-sugars). Those with the opposite configuration are called L-sugars.

Chiral compounds and optical activity are discussed in Section 3.3.2.1

Many carbohydrates exist in equilibrium between open chain and 5- or 6-membered cyclic forms. The cyclic pyranoses (5-ring) or hexanoses (6-ring) are produced from an intramolecular cyclisation of an alcohol (ROH) group onto an aldehyde (RCHO) leading to a hemiacetal, RCH(OH)OR (Section 8.3.5). These are most commonly drawn as *Haworth projections* in which the ring is drawn as a hexagon and vertical lines attach substituents.

Formation of hemiacetals and acetals is discussed in Section 8.3.5.2

Example: D-*glucose*

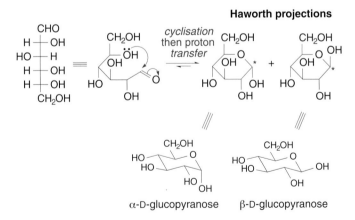

For conformations of cycloalkanes, see Sections 3.2.3 and 3.2.4

α-D-glucopyranose   β-D-glucopyranose

The cyclisation produces two diastereoisomers and the new chiral centre (labelled ∗) which is formed is called the *anomeric carbon*. For glucose, when the anomeric carbon has *S* configuration, it is called the α-anomer while the *R* configuration is called the β-anomer.

When α-D-glucopyranose is dissolved in water, the optical rotation value decreases with time until it reaches $+52.7°$. This is because of equilibration between the α- and β-anomers in a process called *mutarotation*. The β-anomer undergoes reversible ring opening and re-closure, leading to some of the α-anomer.

Disaccharides contain two monosaccharide units joined together by a *glycoside linkage*. This linkage occurs between the anomeric carbon of one sugar and a hydroxyl group at any position of another sugar. The bridging oxygen atom is part of an acetal, $RCH(OR)_2$, and so reaction with aqueous acid (or an appropriate enzyme) leads to hydrolysis and the formation of the component sugars.

Assignment of *R* and *S* configuration is discussed in Section 3.3.2.2

For hydrolysis of acetals see Section 8.3.5.2

an α-1,4'-glycoside linkage

α- and β-D-glucopyranose

maltose
(formed from two D-glucopyranoses)

The wavy line indicates a mixture of anomers

Polysaccharides are polymers of monosaccharides joined by glycoside linkages. The three most abundant natural polysaccharides are cellulose, starch and glycogen, and these are derived from glucose.

11.2 Lipids

Lipids are naturally occurring organic molecules found in cells and tissues, which are soluble in organic solvents but insoluble in water. They are defined by their (organic) solubility rather than by their structure. Lipids can be sub-divided into fats and waxes, and steroids.

11.2.1 Waxes, fats and oils

Natural waxes are esters of long-chain carboxylic acids (RCO_2H) with long-chain alcohols (ROH). The carboxylic acid and alcohol sidechains are usually saturated and contain an even number of carbon atoms.

Saturated compounds contain the maximum number of hydrogen atoms per carbon (Section 2.1)

Example: Beeswax

$$H_3C-(H_2C)_x-C(=O)-O-(CH_2)_y-CH_3 \qquad x = 24, 26; y = 29, 31$$

A hydrolysis reaction is a reaction involving water

Animal fats are solids (e.g. butter), whereas vegetable oils are liquids (e.g. olive oil). They are both triesters of glycerol and are called *triglycerides*. On hydrolysis, they are converted into glycerol and three carboxylic acids, which are called fatty acids.

fat or oil
(triester)

glycerol

fatty acids

Alkyl groups are introduced in Section 2.2

The fatty acids usually contain between 12 and 20 carbon atoms and the alkyl side chains can be saturated or unsaturated (i.e. can contain alkene double bonds, C=C). Vegetable oils contain a higher proportion of unsaturated fatty acids than animal fats. The shape of the alkenes (which usually have the *cis-* configuration) prevent the molecules packing closely together. This makes it harder for the molecules to crystallise, which lowers the melting point. This explains why (unsaturated) vegetable oils are liquids.

For *cis-* and *trans*-isomerism of alkenes see Section 3.3.1.1

- Catalytic hydrogenation of the alkene C=C bonds in vegetable oils (known as hardening) is carried out industrially to produce margarine.
- Soap, which is a mixture of sodium and potassium salts of fatty acids, is produced industrially by hydrolysis (saponification) of animal fat using aqueous sodium hydroxide.

11.2.2 Steroids

Naturally occurring steroids exert a variety of physiological activities and many act as hormones (i.e. chemical messengers secreted by glands in the body). They all contain one cyclopentane and three cyclohexane rings linked together: the four rings are labelled A–D.

Cyclopentane is introduced in Section 3.2.3

Example

testosterone
(male sex hormone)

For conformations of cyclohexane, see Section 3.2.4

All three cyclohexane rings can adopt strain-free chair conformations, in which the small groups (e.g. hydrogen atoms) at the ring junctions adopt axial positions. Therefore, most steroids have the 'all *trans-*' stereochemistry.

The axial hydrogen atoms at the ring junctions have a *trans*- relationship

Two groups on the opposite sides of the molecule have *trans*-stereochemistry

11.3 Amino acids, peptides and proteins

Peptides and proteins are composed of amino acids ($H_2NCH(R)CO_2H$) linked by amide (or peptide) bonds, RCO–NHR. There are 20 common amino acids, which are naturally occurring, and these all contain an amine group at the α-carbon of a carboxylic acid. All of the 20 most common amino acids contain primary amine groups (RNH_2), except proline, which is a cyclic amino acid bearing a secondary amine (R_2NH).

An α-carbon atom is adjacent to a C=O bond

Primary and secondary amines are introduced in Section 2.1

R = CH(OH)Me, threonine
R = CH₂SH, cysteine
R = CH₂CH₂SMe, methionine
R = CH₂CO₂H, aspartic acid
R = CH₂CH₂CO₂H, glutamic acid
R = CH₂CONH₂, asparagine
R = CH₂CH₂CONH₂, glutamine
R = (CH₂)₄NH₂, lysine
R = (CH₂)₃NHC(=NH)NH₂, arginine
R = CH₂Ph, phenylalanine

R = H, glycine
R = Me, alanine
R = CHMe₂, valine
R = CH₂CHMe₂, leucine
R = CH(Me)Et, isoleucine
R = CH₂OH, serine

Arginine contains a guanidine group in the side-chain, $-NHC(=NH)NH_2$

Methionine contains a sulfide or thioether (R_2S) in the side chain

Cysteine contains a thiol (RSH) in the side chain

, tryptophan   , tyrosine

Tryptophan contains an aromatic heterocycle, called indole, in the side chain

, histidine   , proline

Histidine contains an aromatic heterocycle, called imidazole, in the side chain

The primary amino acids $H_2NCH(R)CO_2H$ differ in the nature of the alkyl side chain (R). All of the 20 common amino acids except glycine (R = H) are chiral, and only one enantiomer is prepared in nature. Most amino acids found in nature have the L-configuration (because of their stereochemical similarity to L-glyceraldehyde). All of the 20 common amino acids have the *S*-configuration except for cysteine.

Assigning *R* or *S* configuration is introduced in Section 3.3.2.2

Fischer projections

Fischer projections are discussed in Section 3.3.2.6

L-alanine
S-configuration

L-glyceraldehyde

Use of solid and dashed lines is introduced in Section 3.3.2

Amino acids exist as *zwitterions* as they contain both a positive and a negative charge within the same molecule. They are *amphoteric*, as they can react with acid to gain a proton, or with base to lose a proton.

low pH   neutral zwitterion   high pH

Amines are good bases (Section 1.7.2)

The pH at which the amino acid exists primarily as the neutral zwitterion is called the *isoelectric point*.

The side chains (R) can be divided into neutral, acidic or basic side chains. Those with alkyl or aryl side chains are neutral, those with amine (or related) side chains are basic, while those with carboxylic acid side chains are acidic.

Reactions of amides are discussed in Section 9.8

Amino acids join together to form the (amide or) *peptide bond* (−NH−CO−) of peptides and proteins. Dipeptides and tripeptides are formed, for example, from combination of two and three amino acids, respectively. The individual amino acids within the peptide are called *residues* and whereas polypeptides usually contain less than 50 residues, proteins often contain more than 50 residues.

peptide bond

A dehydration reaction leads to the loss of water

N-terminus   *C*-terminus

a dipeptide

Peptides are written with the *N*-terminus on the left and the *C*-terminus on the right. The repeating sequence of nitrogen, α-carbon and carbonyl groups is called the *peptide backbone*. The amino acid sequence determines the structure of the peptide/protein. As the peptide/protein size increases, so does the number of possible amino acid combinations. For example, for a protein containing 300 residues there are 20^{300} possible amino acid combinations (as there are 20 common amino acids).

The shape of peptides/proteins is crucial for their biological activity. Although the side-chains of the residues are free to rotate, this is not possible for the peptide bonds. This is because the nitrogen lone pair (of the peptide bond) is conjugated with the carbonyl group (C=O). Therefore, the C−N bond has partial double-bond character and this slows down the rate of rotation around the C−N bond, which helps to make peptides/proteins relatively rigid.

Conjugation and resonance are discussed in Section 1.6.3

Conjugation and amides is discussed in Section 9.3.1

The sequence of amino acids in a protein is called the *primary structure* while the localised spatial arrangement of amino acid segments is called the *secondary structure*. The secondary structure results from the rigidity of the amide bond and any other non-covalent interactions (e.g. hydrogen bonding) of the side-chains. Secondary structures include the *α-helix* and the *β-pleated sheet*. The *tertiary structure* refers to the way in which the entire protein is folded into a 3-dimensional shape, and the *quaternary structure* refers to the way in which proteins come together to form aggregates.

Hydrogen bonding is introduced in Section 1.1

Large proteins, which act as catalysts for biological reactions, are called *enzymes*. The tertiary structure of enzymes usually produces 3-dimensional pockets called *active sites*. The size and shape of the active site is specific for only a certain type of substrate, which is selectively converted into the product by the enzyme. This is often compared to a key fitting a lock (*the lock and key model*). The catalytic activity of the enzyme is destroyed by *denaturation*, which is the breakdown of the tertiary structure (i.e. the protein unfolds). This can be caused by a change in temperature or pH.

11.4 Nucleic acids

The nucleic acids DNA (deoxyribonucleic acid) and RNA (ribonucleic acid) carry the cell's genetic information. Indeed, DNA contains all the information needed for the survival of the cell.

Sugars are discussed in Section 11.1

- Both DNA and RNA are composed of phosphoric acid (H_3PO_4), a sugar, and several heterocyclic organic bases. DNA contains the sugar deoxyribose and RNA contains ribose. The bases adenine (A), guanine (G), cytosine (C) and thymine (T) are present in DNA, while adenine (A), guanine (G), cytosine (C) and uracil (U) are present in RNA.

A heterocycle is a cyclic compound in which at least one atom is not carbon

deoxyribose   ribose

adenine   guanine

purine bases

cytosine   uracil   thymine

pyrimidine bases

- Both DNA and RNA are polymers of *nucleotides* (phosphate–sugar–base), which are formed from *nucleosides* (sugar–base) and phosphoric acid (H_3PO_4). However, the polymer chain of DNA is much larger than that of RNA.

Hemiacetals are discussed in Section 8.3.5.2

Alcohols are introduced in Section 2.1

A polymer contains a repeating sequence of smaller structural units (see Section 11.5)

The structures of both DNA and RNA depend on the sequence of the nucleotides (i.e. the bases). Watson and Crick showed that DNA is a *double helix* composed of two strands with complementary bases, which hydrogen bond to one another. A and T form strong bonds to one another, as does C and G.

Hydrogen bonding is introduced in Section 1.1

RNA is formed by *transcription* of DNA. On cell division, the two chains of the helix unwind and each strand is used as a template for the construction of an RNA molecule. The complementary bases pair up and the completed RNA (which corresponds to only a section of the DNA) then unwinds from the DNA and travels to the nucleus. Unlike DNA, RNA remains a single strand of nucleotides.

11.5 Synthetic polymers

A *polymer* is a large molecule made up of a repeating sequence of smaller units called *monomers*. Naturally occurring polymers include DNA and also cellulose, which is composed of repeating glucose units (Sections 11.1 and 11.4). Synthetic

Proteins are polymers of amino acids (Section 11.3)

polymers, which are made on a large scale in industry, have found a variety of important applications, e.g. adhesives, paints and plastics.

Polymers can be divided into *addition* (or chain-growth) polymers, formed on simple addition of monomers, or *condensation* (or step-growth) polymers, formed on addition of monomers and elimination of a by-product such as water.

In a condensation reaction two molecules react to form a product together with water

11.5.1 Addition polymers

Addition polymers can be formed by reaction of an alkene (e.g. RCH=CHR) with a radical, cation or anion initiator.

Alkenes are discussed in Chapter 6

- *Radical polymerisation* is the most important method and this often employs a peroxide (ROOR) initiator containing a weak oxygen–oxygen bond. On homolysis (using heat or UV radiation) the resulting alkoxyl radical (RO$^\bullet$) adds regioselectively to the least hindered end of the alkene to form a carbon-centred radical. This radical then adds to the least hindered end of another molecule of the alkene to build the polymer by a chain reaction. The polymerisation is terminated by, for example, the coupling of two radicals.

Regioselective reactions are introduced in Section 4.8

For a related radical chain reaction, see Section 6.2.2.1

Notice the use of single-headed (fish-hook) curly arrows (Section 4.1)

Addition of R$^\bullet$ to the least hindered end of the C=C bond produces the most stable carbon-centred radical (Section 4.3.1)

- *Cationic polymerisation* employs a strong protic or Lewis acid initiator. A proton, for example, adds to an electron-rich alkene to form the most stable carbocation (below, a secondary carbocation is formed in preference to a primary carbocation). The carbocation then adds to another electron-rich alkene to build the polymer chain. The polymerisation is terminated by, for example, deprotonation.

Lewis acids are discussed in Section 1.7.3

Carbocation stability is discussed in Section 4.3.1

For a related reaction, namely addition of HBr to an alkene, see Section 6.2.2.1

In reality, H$^+$ does not just drop off the cation, but a base is required

Nucleophiles are introduced in
Section 4.2.1.1

The Michael reaction is discussed
in Section 8.5.4

- *Anionic polymerisation* employs nucleophiles such as alkyllithiums (RLi), alkoxide (RO⁻) or the hydroxide ion (HO⁻) as the initiator. The hydroxide ion, for example, adds to an electron-poor alkene to form the most stable carbanion (in a Michael-type reaction). The carbanion then adds to another electron-poor alkene to build the polymer chain. The polymerisation is terminated by, for example, protonation.

R = electron-withdrawing group (e.g. CO_2Et)

Radical polymerisation of alkenes often leads to chain branching and the formation of nonlinear polymers. In addition, the chiral centres on the backbone of polymers are usually formed randomly to give *atactic* polymers.

Chiral centres, including use of
solid and dashed lines to show 3-D
structure, are discussed in Sections
3.3.2 and 3.3.2.1

The formation of linear *isotactic* or *syndiotactic* polymers can be achieved by metal catalysed polymerisation. This employs *Ziegler-Natta catalysts*, made from triethylaluminium (Et_3Al) and titanium tetrachloride ($TiCl_4$), which react with alkenes by a complex mechanism. Polymerisation of ethylene (ethene, $CH_2{=}CH_2$) leads to the formation of (linear) high-density polyethylene, which is of greater strength than the (branched) low-density polyethylene produced on radical polymerisation.

Copolymers are formed when two or more monomers polymerise to give a single polymer. These polymers often exhibit properties different from that of *homopolymers* (produced from only one alkene).

11.5.2 Condensation polymers

Condensation polymers are formed from two monomers containing two functional groups. The formation of a bond usually leads to the elimination of a simple by-product (such as water) and this occurs in discrete steps (i.e. not via a chain reaction). The polymer is usually composed of an alternating sequence of the two monomers.

- *Polyamides* (or nylons) are produced from heating dicarboxylic acids with diamines. The amine acts as the nucleophile and reacts with the carboxylic acid in a nucleophilic acyl substitution reaction (see Section 9.3).

Reactions of amides are discussed in Section 9.8

A diamine contains two amine groups

- *Polyesters* are produced from heating dicarboxylic acids (or diesters) with diols.

Reactions of esters are discussed in Section 9.7

- *Polyurethanes* are produced from reaction of diols with diisocyanates. Nucleophilic addition of an alcohol to an isocyanate produces the urethane functional group. Although no (small molecule) by-products are produced, the urethane bonds are formed in discrete steps.

An isocyanate has the formula R–N=C=O

Urethanes are also called carbamates

In a proton transfer, H^+ moves from one part of the molecule to another

Worked example

(a) The following questions are based on glutathione, shown below, a naturally occurring antioxidant.

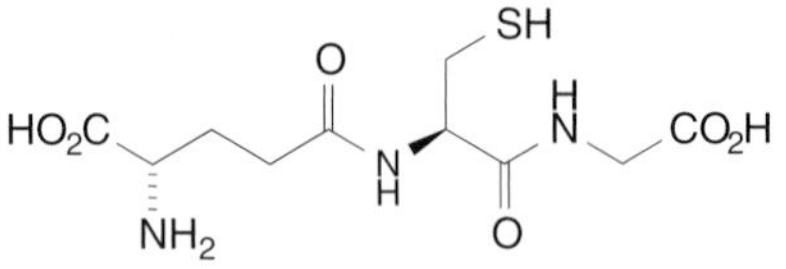

Hint: Consider the amino acid building blocks (Section 11.4)

Hint: Break the amide bonds (Section 11.4)

(i) Glutathione is an example of a tripeptide. Define the term tripeptide.

(ii) Draw the structures of the amino acids that could be combined to form glutathione. Define any chiral centre(s) as *R* or *S*, and name the functional group that contains sulfur.

(b) The following questions relate to lactose, shown below, a sugar found in milk.

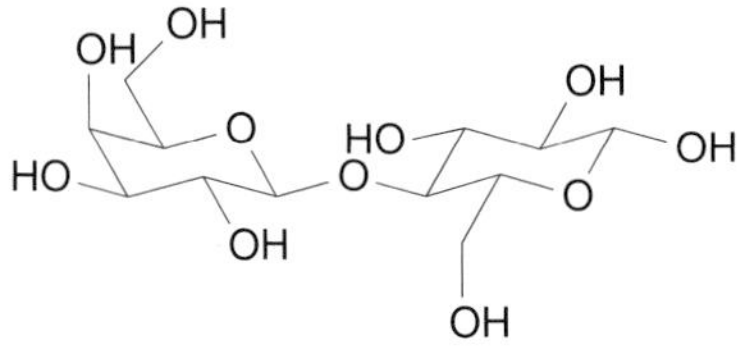

Hint: Consider the sugar building blocks (Section 11.5.2)

(i)   Lactose is an example of a disaccharide. Define the term disaccharide.

(ii) Classify the glycoside linkage as either α or β.

(iii) Lactase is an enzyme that hydrolyses lactose to form two monosaccharides. Draw the structures of the monosaccharides and indicate the anomeric carbons.

(c) The following questions relate to the polymer shown below.

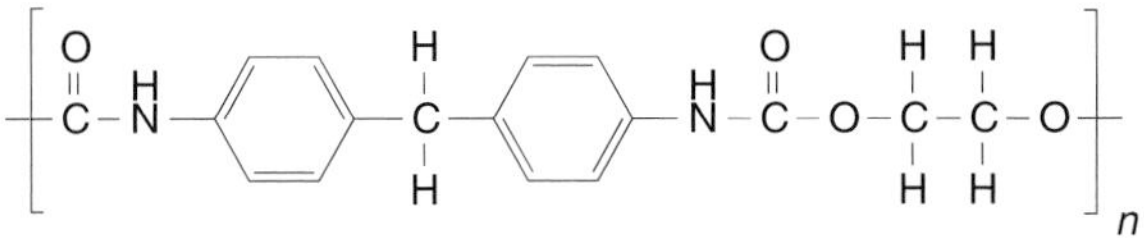

Hint: Consider the repeating functional group (Section 11.5.2)

Peptides are discussed in Section 11.3

For assigning *R* and *S* configuration see Section 3.3.2.2

(i) Is this polymer a polyamide, a polyester or a polyurethane?

(ii) Draw the structures of the component monomers and name the functional groups.

Answer

(a) i. A peptide composed of three amino acids.

ii.

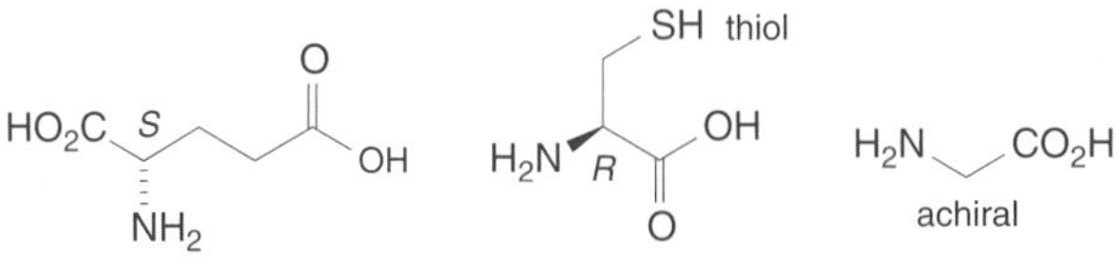

(b) i. A carbohydrate formed from combining two sugars (or monosaccharides).
     ii. The glycoside linkage is β.

     iii.

(c) i.   A polyurethane.
     ii.

diisocyanate   diol

Carbohydrates are discussed in
Section 11.1

Glycoside linkages and anomeric
carbons are discussed in
Section 11.1

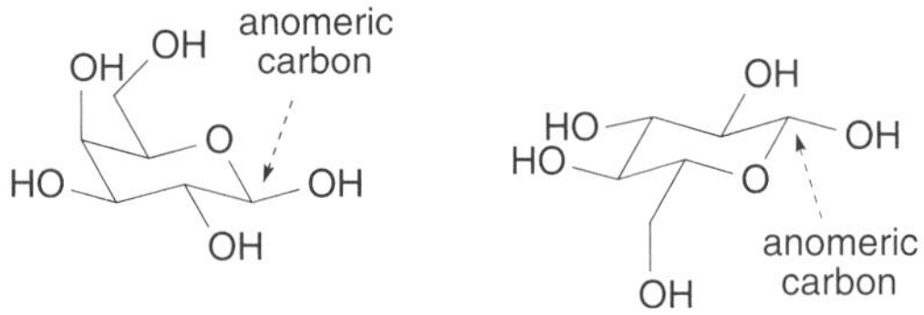

Polyurethanes are discussed in
Section 11.5.2

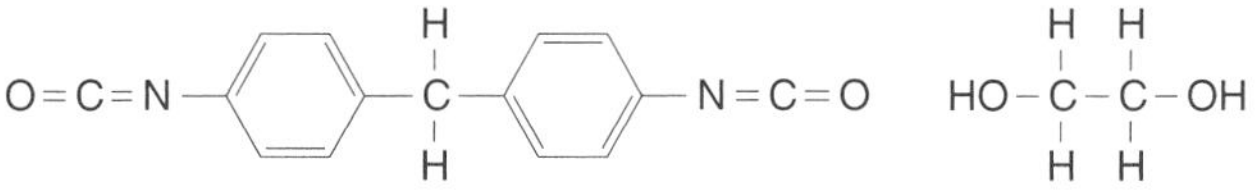

Problems

1. Draw structures of the two possible dipeptides, which can be formed by
    joining (S)-alanine to glycine.

2. The following questions relate to the disaccharide cellobiose **A**.

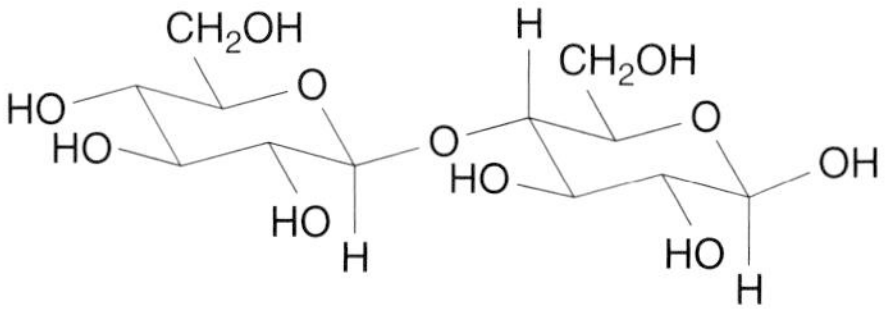

A

    (a) Classify the glycoside linkage as either α or β.
    (b) Name the monosaccharide(s) formed when **A** is hydrolysed in aqueous
        acid.
    (c) Is the monosaccharide(s) formed on hydrolysis of **A** an aldohexose or a
        ketohexose?

3. Are the OH and Me groups in deoxycholic acid **B** (a bile acid) axial or
    equatorial?

B

Section 11.3

Section 11.1

Section 11.2.2

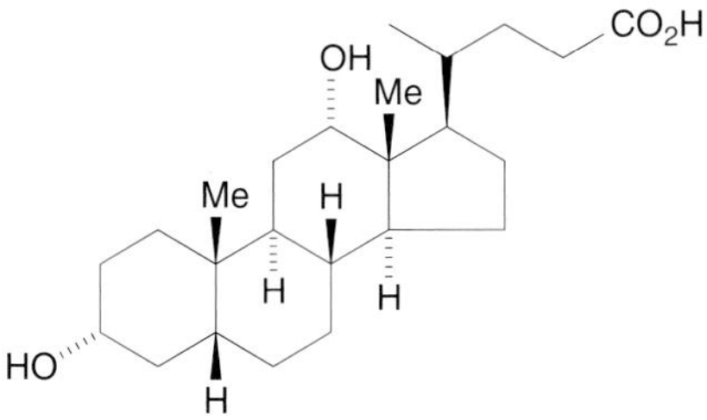

Sections 11.4 and 8.4.1

4. How can tautomerism explain why nucleic acids, such as uracil **C** and guanine **D**, are aromatic even though this is not indicated in the structures shown below?

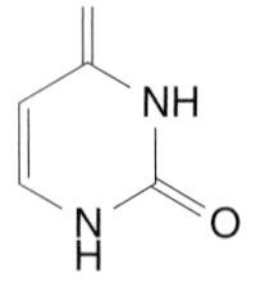

C 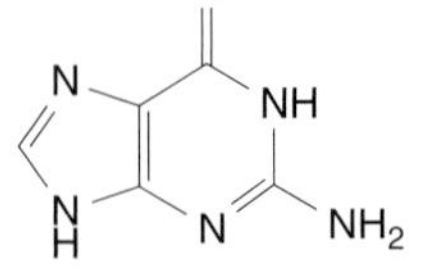   **D**

Section 11.5.1

5. Why is radical polymerisation of vinyl chloride ($H_2C{=}CHCl$) to give poly(vinyl chloride) described as showing a marked preference for head-to-tail addition?

Section 11.5.1

6. Draw the structures of the polymers formed from the following reactions?

   (a)

   (b)

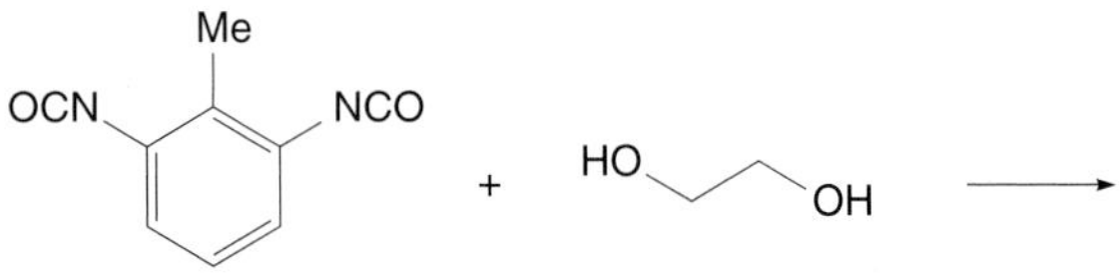

Section 11.4

7. (a) In a DNA double helix, why doesn't A form two hydrogen bonds (out of three possible) with C?

Sections 11.4 and 1.6.3

   (b) In DNA, a guanine residue reacts with electrophiles predominantly at the 7 and 3 positions of the ring system (see below). Suggest an explanation for this.

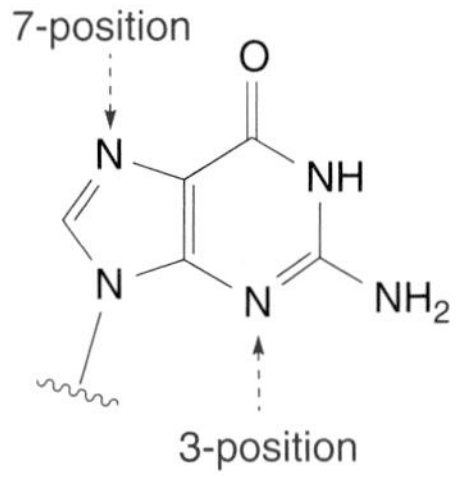

7-position

3-position

Appendix 1

Bond dissociation enthalpies

BDE = approximate bond dissociation enthalpies in kJ mol^{-1}

Bond	H–H	H–F	H–Cl	H–Br	H–I	H–OH	H–NH$_2$
BDE	436	570	431	366	298	464	391

Bond	H$_3$C–H	H$_3$C–Cl	H$_3$C–Br	H$_3$C–I	H$_3$C–OH
BDE	440	356	297	239	389

Bond	C–Cl	C–Br	C–I
Mean BDE	346	290	228

Bond	O–H	R$_3$C–H	N–H
Mean BDE	464	415	391

Bond	C–C	C=C	C≡C
Mean BDE	347	612	838

Bond	C–O	C=O	C–N	C=N	C≡N
Mean BDE	358	742	286	615	887

Bond	B–O	P–O	S–O	P=O	S=O
Mean BDE	515	380	365	510	520

Bond	C–B	P–Cl	P–Br	S–Cl
Mean BDE	395	326	270	271

Keynotes in Organic Chemistry, Second Edition. Andrew F. Parsons.
© 2014 John Wiley & Sons, Ltd. Published 2014 by John Wiley & Sons, Ltd.

Appendix 2

Bond lengths

Bond	H–H	H–F	H–Cl	H–Br	H–I	H–OH	H–NH$_2$
Bond length (Å)	0.74	0.92	1.28	1.42	1.61	0.96	1.01

Bond	C–H	C–F	C–Cl	C–Br	C–I
Mean bond length (Å)	1.09	1.39	1.79	1.94	2.13

Bond	C–C	C=C	C≡C
Mean bond length (Å)	1.53	1.34	1.20

Bond	C–O	C=O	C–N	C=N	C≡N
Mean bond length (Å)	1.42	1.21	1.46	1.21	1.16

Bond	HO–OH	O=O	C≡O
Bond length (Å)	1.48	1.21	1.13

Bond	Cl–Cl	Br–Br	I–I
Bond length (Å)	1.99	2.28	2.67

Keynotes in Organic Chemistry, Second Edition. Andrew F. Parsons.
© 2014 John Wiley & Sons, Ltd. Published 2014 by John Wiley & Sons, Ltd.

Appendix 3

Approximate pK_a values (relative to water)

Name (functional group)	Acid/formula	Conjugate base	Typical pK_a
sulfuric acid	H_2SO_4	HSO_4^-	−3
alkyloxonium ion	ROH_2^+	ROH	−2.4
hydronium ion	H_3O^+	H_2O	−1.7
nitric acid	HNO_3	NO_3^-	−1.4
carboxylic acid	RCO_2H	RCO_2^-	4−5
pyridinium ion	$C_5H_5NH^+$	C_5H_5N	5
carbonic acid	H_2CO_3	HCO_3^-	6.4
ammonium ion	H_4N^+	H_3N	9.3
phenol	C_6H_5OH	$C_6H_5O^-$	10
triethylammonium ion	Et_3NH^+	Et_3N	10.7
β-ketoester	$RCOCH_2CO_2R$	$RCOCH^-CO_2R$	11
water	H_2O	HO^-	15.7
amide	$RNHCOR$	RN^-COR	16
alcohol	ROH	RO^-	16−17
ketone	$RCOCHR_2$	$RCOC^-R_2$	19
ester	R_2CHCO_2R	$R_2C^-CO_2R$	25
alkyne	$RC{\equiv}CH$	$RC{\equiv}C^-$	25
hydrogen	H_2	H^-	35
ammonia	H_3N	H_2N^-	38
benzylic alkyl	$ArCH_3$	$ArCH_2^-$	42
alkene	$R_2C{=}CHR$	$R_2C{=}C^-R$	44
alkane	R_3CH	R_3C^-	≥50

Carboxylic acid	Conjugate base	Typical pK_a
$CH_3CH_2CO_2H$	$CH_3CH_2CO_2^-$	4.9
CH_3CO_2H	$CH_3CO_2^-$	4.8
HCO_2H	HCO_2^-	3.8
$BrCH_2CO_2H$	$BrCH_2CO_2^-$	2.9
FCH_2CO_2H	$FCH_2CO_2^-$	2.7

Keynotes in Organic Chemistry, Second Edition. Andrew F. Parsons.
© 2014 John Wiley & Sons, Ltd. Published 2014 by John Wiley & Sons, Ltd.

Appendix 4

Useful abbreviations

Ac	acetyl [$CH_3C(O)-$]
aq.	aqueous
Ar	aryl
9-BBN	9-borabicyclo[3.3.1]nonane
br	broad (absorption in spectroscopy)
Bn	benzyl ($PhCH_2-$)
Bu	butyl ($CH_3CH_2CH_2CH_2-$)
^{i}Bu	isobutyl or 2-methylpropyl [$(CH_3)_2CHCH_2-$]
^{s}Bu	*sec*-butyl or 1-methylpropyl [$CH_3CH_2CH(CH_3)-$]
^{t}Bu	*tert*-butyl or 1,1-dimethylethyl [$(CH_3)_3C-$]
Bz	benzoyl [$PhC(O)-$]
CI	chemical ionisation (in mass spectrometry)
d	doublet (in ^1H NMR spectroscopy)
DIBAL-H	diisobutylaluminium hydride ($^iBu_2AlH$)
DMF	dimethylformamide [$(CH_3)_2NCHO$]
DMSO	dimethyl sulfoxide [$CH_3S(O)CH_3$]
E	electrophile
e^-	electron
ee	enantiomeric excess (0% ee = racemisation)
EI	electron impact (in mass spectrometry)
Et	ethyl (CH_3CH_2-)
h	hours
HOMO	highest occupied molecular orbital
hv	radiation energy
IR	infrared
J	coupling constant/Hz (in ^1H NMR spectroscopy)
k	rate constant
LUMO	lowest unoccupied molecular orbital
*m*CPBA	*meta*-chloroperbenzoic acid ($3\text{-}ClC_6H_4CO_3H$)
Me	methyl
mp	melting point
MS	mass spectrometry

Keynotes in Organic Chemistry, Second Edition. Andrew F. Parsons.
© 2014 John Wiley & Sons, Ltd. Published 2014 by John Wiley & Sons, Ltd.

m/z	mass:charge ratio
NBS	*N*-bromosuccinimide
NMR	nuclear magnetic resonance
Nu	nucleophile
m	multiplet (in ^{1}H NMR spectroscopy)
PCC	pyridinium chlorochromate ($C_5H_5NH^+$ $ClCrO_3^-$)
Ph	phenyl (C_6H_5-)
Pr	propyl ($CH_3CH_2CH_2-$)
^{i}Pr	isopropyl or 1-methylethyl [($CH_3CH(CH_3)-$]
Py	pyridine (C_5H_5N)
q	quartet (in ^{1}H NMR spectroscopy)
R	alkyl group (methyl, ethyl, propyl etc)
R,S	configuration about a stereogenic centre
s	singlet (in ^{1}H NMR spectroscopy)
THF	tetrahydrofuran
TMS	tetramethylsilane (Me_4Si)
t	triplet (in ^{1}H NMR spectroscopy)
Ts	4-toluenesulfonyl, (4-$CH_3C_6H_4SO_2-$)
UV	ultraviolet
X	halogen atom

Appendix 5

Infrared absorptions

Functional group	Typical wavenumber/cm^{-1}
alcohol, O–H	3600–3200
amine, N–H	3500–3300
alkane, C–H	3000–2850
nitrile, C≡N	2250–2200
alkyne, C≡C	2200–2100
carbonyl, C=O	1820–1650
imine, C=N	1690–1640
alkene, C=C	1680–1620
benzene, CC	1600–1500
ether, C–O	1250–1050
amine, C–N	1250–1020
chloroalkane, R–Cl	800–600
bromoalkane, R–Br	750–500
iodoalkane, R–I	~500

Functional group	Typical wavenumber of C=O stretch/cm^{-1}
acyl chloride, RCOCl	1820
acid anhydride, RCO$_2$COR	1800
ester, RCO$_2$R	1740
aldehyde, RCHO	1730
ketone, RCOR	1715
carboxylic acid, RCO$_2$H	1710
amide, RCONH$_2$	1650

Keynotes in Organic Chemistry, Second Edition. Andrew F. Parsons.
© 2014 John Wiley & Sons, Ltd. Published 2014 by John Wiley & Sons, Ltd.

Approximate NMR chemical shifts

^{1}H NMR chemical shifts

Methyl (CH$_3$) hydrogens

Functional group	Typical chemical shift (δ) of methyl hydrogens/ppm
C–CH$_3$	0.9
O–C–CH$_3$	1.3
C=C–CH$_3$	1.6
R–O–C(=O)–CH$_3$	2.0
R–C(=O)–CH$_3$	2.2
Ar–CH$_3$	2.3
N–CH$_3$	2.3
Ar–C(=O)–CH$_3$	2.5
R–C(=O)N–CH$_3$	2.9
R–O–CH$_3$	3.3
R–C(=O)O–CH$_3$	3.7
Ar–O–CH$_3$	3.8

Methylene (CH$_2$) hydrogens

Functional group	Typical chemical shift (δ) of methylene hydrogens/ppm
C–CH$_2$–C	1.4
O–C–CH$_2$–C	1.9
R–O–C(=O)–CH$_2$–C	2.2
C=C–CH$_2$–C	2.3
R–C(=O)–CH$_2$–C	2.4
N–CH$_2$–C	2.5
Ar–CH$_2$–C	2.7
Ar–C(=O)–CH$_2$–C	2.8
R–O–CH$_2$–C	3.4
Br–CH$_2$–C	3.4
Cl–CH$_2$–C	3.5
HO–CH$_2$–C	3.6
Ar–O–CH$_2$–C	4.0
R–C(=O)O–CH$_2$–C	4.1

Keynotes in Organic Chemistry, Second Edition. Andrew F. Parsons.
© 2014 John Wiley & Sons, Ltd. Published 2014 by John Wiley & Sons, Ltd.

Methine (CH) hydrogen

Functional group	Typical chemical shift (δ) of methine hydrogens/ppm
C–CH–C	1.5
O–C–CH–C	2.0
R–O–C(=O)–CH–C	2.5
R–C(=O)–CH–C	2.7
N–CH–C	2.8
Ar–CH–C	3.0
Ar–C(=O)–CH–C	3.4
R–O–CH–C	3.7
HO–CH–C	3.7
Br–CH–C	4.3
ArO–CH–C	4.5
R–C(=O)O–CH–C	4.8
Ar–C(=O)O–CH–C	5.1

^{13}C NMR chemical shifts

Functional group	Typical chemical shift (δ) of carbons/ppm
C=O	220–165
substituted benzene	170–110
C=C	150–110
C–N, C–O	80–40
C–Cl, C–Br	75–25
C–C	50–5

Examples: Approximate ^{1}H NMR and ^{13}C NMR chemical shifts

^{1}H NMR chemical shifts in black and ^{13}C NMR shifts in blue

3-Methyl-1-phenylbutan-1-one

Ethyl 2-bromopropanoate

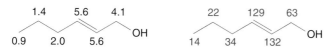

(*E*)-Hex-2-en-1-ol

4-Ethylbenzaldehyde

2-Methoxyethyl butyrate

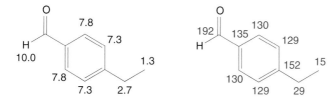

Appendix 7

Reaction summaries

Chapter 5

Keynotes in Organic Chemistry, Second Edition. Andrew F. Parsons.
© 2014 John Wiley & Sons, Ltd. Published 2014 by John Wiley & Sons, Ltd.

Chapter 6

Chapter 7

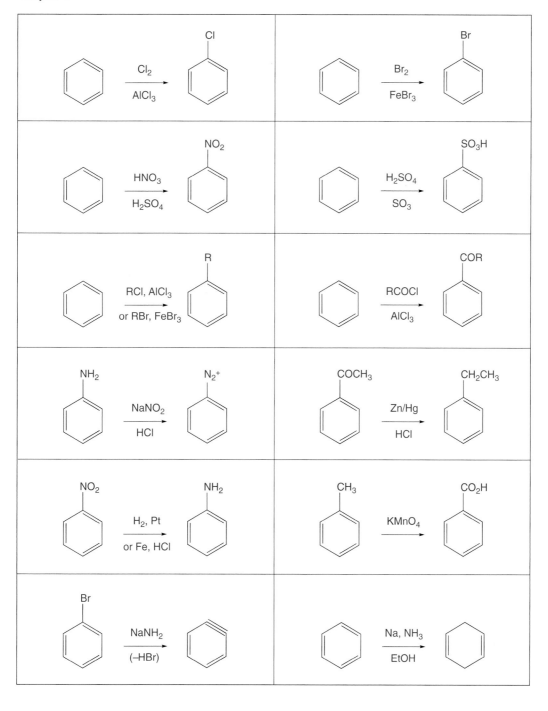

Chapter 8

R–C(=O)–R $\xrightarrow[\text{then } \mathbf{H^+}]{\text{NaBH}_4 \text{ or LiAlH}_4}$ HO–CH(R)(R)	R–C(=O)–R $\xrightarrow[\text{HCN}]{\text{NaCN (catalytic)}}$ HO–C(CN)(R)(R)
R–C(=O)–R $\xrightarrow[\text{then } \mathbf{H^+}]{\text{RLi or RMgX}}$ HO–C(R)(R)(R)	R–C(=O)–R $\xrightarrow[(-\text{Ph}_3\text{P=O})]{\text{Ph}_3\text{P=CHR}}$ CHR=C(R)(R)
R–C(=O)–R $\xrightarrow[(\text{H}^+ \text{ or HO}^- \text{ catalyst})]{\text{H}_2\text{O}}$ HO–C(OH)(R)(R)	R–C(=O)–R $\xrightarrow[\substack{\text{H}^+ \text{ (catalyst)}, \\ -\text{H}_2\text{O}}]{2\text{ROH}}$ RO–C(OR)(R)(R)
R–C(=O)–R $\xrightarrow[\substack{\text{H}^+ \text{ (catalyst)}, \\ -\text{H}_2\text{O}}]{2\text{RSH}}$ RS–C(SR)(R)(R)	R–C(=O)–R $\xrightarrow[\substack{\text{H}^+ \text{ (catalyst)}, \\ -\text{H}_2\text{O}}]{\text{RNH}_2}$ NR=C(R)(R)
R–C(=O)–CH$_3$ $\xrightarrow[\substack{\text{H}^+ \text{ (catalyst)}, \\ -\text{HOH}}]{\text{R}_2\text{NH}}$ NR$_2$–C(R)=CH$_2$	R–C(=O)–CH$_3$ $\xrightarrow[\substack{\text{H}^+ \text{ (catalyst)}, \\ -\text{HCl}}]{\text{Cl}_2}$ R–C(=O)–CH$_2$Cl
R–C(=O)–CH$_3$ $\xrightarrow[\text{then } \mathbf{H^+}]{\text{I}_2, \text{HO}^-}$ R–CO$_2$H + CHI$_3$	EtO$_2$C–CH$_2$–CO$_2$Et $\xrightarrow[\text{then R–Br}]{\text{EtO}^-}$ EtO$_2$C–CHR–CO$_2$Et
2 H–C(=O)–CH$_3$ $\xrightarrow[(-\text{H}_2\text{O})]{\text{HO}^-, \text{ heat}}$ H–C(=O)–CH=CH–CH$_3$	R–C(=O)–CH=CH–R $\xrightarrow[\text{then } \mathbf{H^+}]{\text{R}_2\text{CuLi}}$ R–C(=O)–CH$_2$–CHR–R

Chapter 9

$R\text{—COOH} \xrightarrow[\text{or PCl}_5]{\text{SOCl}_2} R\text{—COCl}$	$R\text{—COOH} \xrightarrow[\substack{\text{H}^+ \text{ (catalyst),} \\ -\text{HOH}}]{\text{ROH}} R\text{—COOR}$
$R\text{—COCl} \xrightarrow[(-\text{HCl})]{\text{H}_2\text{O}} R\text{—COOH}$	$R\text{—COCl} \xrightarrow[(-\text{HCl})]{\text{NH}_3} R\text{—CONH}_2$
$R\text{—CO—OCOR} \xrightarrow{\text{H}_2\text{O}} R\text{—COOH} + \text{RCO}_2\text{H}$	$R\text{—CO—OCOR} \xrightarrow{\text{ROH}} R\text{—COOR} + \text{RCO}_2\text{H}$
$R\text{—COOR} \xrightarrow[\text{then }\mathbf{H^+}]{\text{LiAlH}_4} R\text{—CH}_2\text{OH}$	$R\text{—COOR} \xrightarrow[\text{then }\mathbf{H^+}]{\text{RMgX}} R\text{—CR}_2\text{OH}$
$R\text{—COOH} \xrightarrow[\text{then }\mathbf{H^+}]{\text{LiAlH}_4} R\text{—CH}_2\text{OH}$	$R\text{—COOR} \xrightarrow[(-\text{ROH})]{\text{H}^+, \text{H}_2\text{O}} R\text{—COOH}$
$R\text{—CONH}_2 \xrightarrow[\text{H}_2\text{O}]{\text{H}^+} R\text{—COOH}$	$R\text{—C}\equiv\text{N} \xrightarrow[\text{H}_2\text{O}]{\text{H}^+} R\text{—COOH}$
$\text{HO—CO—CHR}_2 \xrightarrow[\text{then H}_2\text{O}]{\text{Br}_2, \text{PBr}_3} \text{HO—CO—CBrR}_2$	$2\ \text{RO—CO—CH}_3 \xrightarrow[\text{then H}^+]{\text{RO}^-} \text{RO—CO—CH}_2\text{—CO—CH}_3$

Appendix 8

Glossary

α-Carbon	The carbon atom directly attached to the functional group of interest, typically a carbonyl group (C=O)
α-Hydrogen	A hydrogen atom directly bonded to the α-carbon
Achiral	An achiral molecule has a superimposable mirror image
Activating group	A substituent that increases the reactivity of an aromatic ring to electrophilic substitution
Activation energy	The difference in energy between the reactants and the transition state
Acyclic	A molecule, or part of a molecule, whose atoms are not part of a ring
Acyl	A group containing C=O bonded to an alkyl group or an aryl group (e.g. $-COCH_3$, $-COPh$)
Addition reaction	Two groups added to opposite ends of a π bond
Alicyclic	Cyclic compounds that are not aromatic
Aliphatic	Compounds that are not cyclic and not aromatic
Alkyl group	Formed on removal of a hydrogen atom from an alkane (e.g. methane, CH_4, gives methyl, CH_3)
Alkylation	A reaction that adds an alkyl group to a reactant
Angle strain	Strain due to deviation from one or more ideal bond angles
Anhydrous	Without water
Anti	When two atoms or groups point in opposite directions; when the atoms or groups lie in the same plane they are *antiperiplanar*
Anti addition	An addition reaction in which two substituents add to the opposite sides of the molecule

Keynotes in Organic Chemistry, Second Edition. Andrew F. Parsons.
© 2014 John Wiley & Sons, Ltd. Published 2014 by John Wiley & Sons, Ltd.

Anti elimination	An elimination reaction in which two substituents are eliminated from opposite sides of the molecule
Antibonding molecular orbital	A molecular orbital formed by out-of-phase overlap of atomic orbitals
Aprotic solvent	A solvent that is not a hydrogen bond donor
Aromatic	Cyclic compounds such as benzene and related ring systems
Aryl	A phenyl group ($-C_6H_5$), or a substituted phenyl group
Asymmetric centre	An atom bonded to four different atoms or groups
Basicity	The tendency of a molecule to share its electrons with a proton
β-Elimination	A reaction in which two atoms/groups on adjacent atoms are lost to form a π bond (e.g. elimination of HBr from $BrCH_2CH_3$ to form $H_2C{=}CH_2$)
Bimolecular	A reaction whose rate depends on the concentration of two reactants (e.g. an S_N2 reaction)
Bond angle	The angle formed by three contiguous bonded atoms
Bonding molecular orbital	A molecular orbital formed by in-phase overlap of atomic orbitals
Carbanion	A species with a negative charge on carbon
Carbene	A species with a nonbonded pair of electrons on carbon and an empty orbital (e.g. $H_2C{:}$)
Carbocation	A species with a positive charge on carbon
Catalyst	A species that increases the rate at which a reaction occurs without being consumed in the reaction
Catalytic	The reactant is a catalyst, and is present in a small amount
Chemoselectivity	Preferential reaction of one functional group in the presence of others
Chiral	A chiral molecule has a nonsuperimposable mirror image
Concerted reaction	A single-step reaction in which reactants are directly converted into products without the involvement of any intermediates
Configuration	The arrangement of atoms or groups in a molecule to give configurational isomers that cannot be interconverted without breaking a bond

Conformation	The three-dimensional arrangement of atoms that result from rotation of a single bond
Conformer	A specific conformation of a molecule that is relatively stable
Conjugate acid	The acid formed on protonation of a base
Conjugate base	The base formed on deprotonation of an acid
Conjugation	Stability associated with molecules containing alternating single and double bonds, due to overlapping p orbitals and electron delocalisation
Constitutional isomers	Molecules that have the same molecular formula but differ in the way their atoms are connected
Coordinate (dative) bond	A covalent bond where one of the atoms provides both electrons
Covalent bond	A bond formed as a result of sharing electrons
Cyclisation	A reaction leading to the formation of a ring
Cycloaddition	An addition reaction that forms a ring
Deactivating group	A substituent that decreases the reactivity of an aromatic ring to electrophilic substitution
Decarboxylation	Loss of carbon dioxide
Dehydration	Loss of water
Delocalisation	When lone pairs or electrons in π bonds are spread over several atoms
Deprotonate	To remove a proton (H^+)
Diastereoisomer	Stereoisomers that are not mirror images of each other
Dielectric constant	A measure of how well a solvent can insulate opposite charges from one another (polar solvents have relatively high dielectric constants)
Dihedral angle	The angle between two planes, where each plane is defined by three atoms
Dipole moment	A measure of the separation of charge in a bond or in a molecule
Disubstituted	Typically an alkene or substituted benzene that contains two groups, or atoms other than hydrogen
Electron donating group	An atom or group that releases electron density to neighbouring atoms
Electron withdrawing group	An atom or group that draws electron density from neighbouring atoms towards itself
Electronegative	An atom that attracts electrons toward itself
Electronic effect	The reactivity of a part of the molecule is affected by electron attraction or repulsion

Electrophile	Electron-deficient reactant that accepts two electrons to form a covalent bond
Electrophilicity	The relative reactivity of an electrophilic reagent
Elimination	A reaction that involves the loss of atoms or molecules from the reactant, typically from adjacent atoms
Enantiomer (optical isomer)	One of a pair of molecules which are mirror images of each other and nonsuperimposable
Endothermic	A reaction in which the enthalpy of the products is greater than the enthalpy of the reactants; the overall standard enthalpy change is positive
Energy profile	A plot of the conversion of reactants into products versus energy (Gibbs free energy or enthalpy)
Enolisation	The conversion of a keto form into an enol form
Enthalpy	The heat given off or the heat absorbed during the course of a reaction
Entropy	A measure of the disorder or randomness in a closed system
Equilibrium constant	The ratio of products to reactants at equilibrium or the ratio of the rate constants for the forward and reverse reactions
Exothermic	A reaction in which the enthalpy of the products is smaller than the enthalpy of the reactants; the overall standard enthalpy change is negative
Formal charge	A positive or negative charge assigned to atoms that have an apparent abnormal number of bonds
Functional group	An atom, or a group of atoms that has similar chemical properties whenever it occurs in different compounds
Geometric isomers	*cis*/*trans* or *E*/*Z* isomers
Heteroaromatic	An aromatic molecule that contains at least one heteroatom as part of the aromatic ring (e.g. pyridine)
Heteroatom	Any atom other than carbon or hydrogen
Heterocycle	A cyclic molecule where at least one atom in the ring is not carbon
Heterolysis	Breaking a bond unevenly, so that both electrons stay with one of the atoms
Homolysis	Breaking a bond evenly, so that each atom gets one electron

Hybridisation	Mixing atomic orbitals to form new hybrid orbitals
Hydration	Addition of water to the reactant
Hydrogen bond	A noncovalent attractive force caused when the partially positive hydrogen of one molecule interacts with the partially negative heteroatom of another molecule
Hydrogenation	Addition of hydrogen
Hydrolysis	A reaction in which water is a reactant
Hydrophilic	A molecule or part of a molecule with high polarity ('water loving')
Hydrophobic	A molecule or part of a molecule with low polarity ('water-fearing')
Hyperconjugation	Donation of electrons from C−H or C−C sigma bonds to an adjacent empty p orbital
Inductive effect	The polarisation of electrons in sigma bonds
Intermediate	A species with a lifetime appreciably longer than a molecular vibration (corresponding to a local potential energy minimum) that is formed from the reactants and reacts further to give (either directly or indirectly) the products of a reaction
Intermolecular	A process that occurs between two or more separate molecules
Intramolecular	A process that occurs within a single molecule
Ionic bond	A bond formed through the attraction of two ions of opposite charges
Isomerisation	The process of converting a molecule into its isomer
Isomers	Non-identical compounds with the same molecular formula
Kinetic control	A reaction in which the product ratio is determined by the rate at which the products are formed
Leaving group	An atom or group (charged or uncharged) that becomes detached from the main part of a reactant, that takes a pair of electrons with it
Lewis acid	Accepts a pair of electrons
Lewis base	Donates a pair of electrons
Lone pair	Two (paired) electrons in the valence shell of a single atom that are not part of a covalent bond
Mechanism	A step-by-step description of the bond changes in a reaction, often shown using curly arrows

Mesomeric effect	Delocalisation of electron density through π bonds
Molecular orbital	An orbital which extends over two or more atoms
Monomer	A repeating unit in a polymer
Nucleophile	Electron-rich reactant that donates two electrons to form a covalent bond
Nucleophilicity	The relative reactivity of a nucleophilic reagent
One equivalent	Amount of a substance that reacts with one mole of another substance
Orbital	The volume of space around the nucleus in which an electron is most likely to be found
Oxidising agent	In a reaction, the reactant that causes the oxidation (and becomes reduced)
Oxidation reaction	A reaction that increases the number of covalent bonds between carbon and a more electronegative atom (e.g. O, N, a halogen atom), and decreases the number of C–H bonds
Pi bond	A covalent bond formed by side-on overlapping between atomic orbitals
pK_a	The tendency of a compound to lose a proton (a quantitative measure of acidity)
Polar bond	A covalent bond formed by the unequal sharing of electrons
Polar reaction	Reaction of a nucleophile with an electrophile
Polarisability	Indicates the ease with which the electron cloud of an atom can be distorted
Polymer	A molecule in which one or more subunits (called monomers) is repeated many times
Protic solvent	A solvent that is a hydrogen bond donor
Racemate	A mixture of equal amounts of a pair of enantiomers
Radical (or free radical)	An atom or molecule containing an unpaired electron
Radical anion	A species with a negative charge and an unpaired electron
Radical cation	A species with a positive charge and an unpaired electron
Rate-determining step	The step in a reaction that has the transition state with the highest energy
Reducing agent	In a reaction, the reactant that causes the reduction (and becomes oxidised)
Reduction reaction	A reaction that decreases the number of covalent bonds between carbon and a more

	electronegative atom (e.g. O, N, a halogen atom), and increases the number of C—H bonds
Regioselective	A reaction that favours bond formation at a particular atom over other possible atoms
Resonance	When a molecule with delocalised electrons is more accurately described by two or more structures
Resonance energy	The extra stability gained by electron delocalisation due to resonance
Resonance hybrid	The actual structure of a compound with delocalised electrons; it is the average of two or more resonance forms
Resonance structure	One of a set of structures that differ only in the distribution of electrons in covalent bonds and lone pairs
Saturated	A compound with no double or triple bond
Sigma bond	A covalent bond formed by head-on overlapping between atomic orbitals
Skeletal structure	Represents the C—C bonds as lines and does not show the C—H bonds
Solvation	The interaction between a solvent and another molecule or ion
Solvolysis	Reaction of the reactant with the solvent
Stereochemistry	The study of the spatial relationship of atoms within a molecule
Stereoisomers	Isomers that differ in the way their atoms are arranged in space (enantiomers and diastereoisomers are stereoisomers)
Stereoselectivity	A reaction that leads to preferential formation of one stereoisomer (enantiomer, diastereoisomer, or alkene isomer) over another
Stereospecific	A reaction in which the stereochemistry of the reactant controls the outcome of the reaction (e.g. an E2 elimination is stereospecific)
Steric effect	Any effect on a molecule or reaction due to the size of atoms
Steric hindrance	Describes bulky groups at the site of a reaction that make it difficult for the reactants to approach each other
Substituent	An atom or group, other than hydrogen, in a molecule
Substitution reaction	A reaction in which an atom or group is replaced by another atom or group

Syn	When two atoms or groups point in the same direction; when the atoms or groups lie in the same plane they are *synperiplanar*
Syn addition	An addition reaction in which two substituents add to the same side of the molecule
Syn elimination	An elimination reaction in which two substituents are eliminated from the same side of the molecule
Tautomers	Rapidly equilibrating isomers that differ in the location of their bonding electrons (e.g. keto and enol forms of an aldehyde)
Thermodynamic control	A reaction in which the product ratio is determined by the relative stability of the products
Torsional strain	On bond rotation, the strain caused by repulsion of electrons when different groups pass one another
Transition state	The highest energy structure along the reaction coordinate between reactants and products for every step of a reaction mechanism
Unimolecular	A reaction whose rate depends on the concentration of one reactant (e.g. an S_N1 reaction)
Unsaturated	A compound with one or more double or triple bonds
Valence electron	An electron in the outermost shell of an atom
Ylide	A compound with both a negative charge and a positive charge on adjacent atoms
Zwitterion	A compound with both a negative charge and a positive charge that are on nonadjacent atoms

Further reading

There are a large number of excellent organic textbooks and websites that can be used to consolidate your understanding of the topics discussed in this book. Some examples are highlighted below.

General chemistry textbook

A. D. Burrows, J. S. Holman, A. F. Parsons, G. M. Pilling and G. J. Price, Chemistry[3], 2nd Edition, Oxford University Press, 2013.

and http://global.oup.com/uk/orc/chemistry/burrows2e/

General organic textbooks

P. Y. Bruice, Organic Chemistry, 7th Edition, Pearson, 2013.
J. Clayden, N. Greeves and S. Warren, Organic Chemistry, 2nd Edition, Oxford University Press, 2012.

and http://global.oup.com/uk/orc/chemistry/clayden2e/

J. McMurry, Organic Chemistry, 7th Edition, Brooks Cole, 2007.

Keynotes in Organic Chemistry, Second Edition. Andrew Parsons.
© 2014 John Wiley & Sons, Ltd. Published 2014 by John Wiley & Sons, Ltd.

Reaction mechanisms

M. G. Moloney, Reaction Mechanisms at a Glance: A Stepwise Approach to Problem-Solving in Organic Chemistry, Wiley-Blackwell, 1999.

More specialist organic texts

See the Oxford chemistry primer series, which includes many relevant titles including the following.

J. Jones, Core Carbonyl Chemistry, Oxford University Press, 1997.
G. M. Hornby and J. M. Peach, Foundations of Organic Chemistry, Oxford University Press, 1997.
M. Sainsbury, Aromatic Chemistry, Oxford University Press, 1992.
S. B. Duckett and B. C. Gilbert, Foundations of Spectroscopy, Oxford University Press, 2000.

Interactive 3D animations

ChemTube3D: http://www.chemtube3d.com/

Outline answers

Chapter 1

1. (a) $+I$ (b) $-I, +M$ (c) $-I, +M$ (d) $-I, +M$
    (e) $-I, +M$ (f) $-I, -M$ (g) $-I, -M$ (h) $-I, -M$

2. (a)

(b) Expect **C** to be the more stable because of the $+M$ effect of the OMe substituent.

3. (a) The carbocation $CH_3OCH_2^+$ is stabilised by the $+M$ effect of the OCH_3 substituent.

    (b) The electron-withdrawing NO_2 ($-I, -M$) group can stabilise the phenoxide anion by delocalisation of the negative charge. The negative charge can be spread on to the NO_2 group at the 4-position.

    (c) On deprotonation of CH_3COCH_3 the anion (called an enolate ion) can be stabilised by delocalisation of the negative charge on to the $C=O$ group. As a consequence, CH_3COCH_3 is more acidic than CH_3CH_3.

    (d) This can be explained by resonance. Drawing a second resonance structure for $CH_2=CH-CN$ shows that the $C-C$ bond in $CH_2=CH-CN$ has partial double bond character (see below). This is not possible for CH_3-CN.

The $+M$ effect of an OR group is illustrated in Section 1.6.3

Resonance forms of nitro-substituted phenoxide ions are discussed in Section 1.7.1

Keynotes in Organic Chemistry, Second Edition. Andrew F. Parsons.
© 2014 John Wiley & Sons, Ltd. Published 2014 by John Wiley & Sons, Ltd.

(e) For $CH_2=CH-CH_2^+$, the carbon atom bearing the positive charge has six valence electrons. It can accept two further electrons to generate an alternative resonance structure.

For $CH_2=CH-NMe_3^+$, the nitrogen atom bearing the positive charge has eight valence electrons. It cannot expand its valence electrons to ten and so an alternative resonance structure cannot be drawn.

4. On deprotonation of cyclopentadiene, an anion with 6π electrons is formed. This anion is stabilised by aromaticity, hence the low pK_a value of cyclopentadiene.

On deprotonation of cycloheptatriene, an anion with 8π electrons is formed. This anion is not stabilised by aromaticity, hence the high pK_a value of cycloheptatriene.

For approximate pK_a values for these hydrogen atoms:

PhOH	9.9
RCO_2H	4–5
$RC\equiv CH$	25
ROH	16–17

For approximate pK_a values of the conjugate acids of these bases:

Guanidine	13
$PrNH_2$	11
Aniline	4.9
Ethanamide	−0.5

5. The most acidic hydrogen atom in each compound is shown in blue.
   (a) $4\text{-}HOC_6H_4CH_3$
   (b) $4\text{-}HOC_6H_4CO_2H$
   (c) $H_2C=CHCH_2CH_2C\equiv CH$
   (d) $HOCH_2CH_2CH_2C\equiv CH$

6. (a) guanidine > 1-aminopropane > aniline > ethanamide
   On protonation of guanidine, the $(H_2N)_2C=NH_2^+$ cation is stabilised by extensive delocalisation of the charge by the electron donating NH_2 (+M) groups (see below). Normally, an sp^3 nitrogen is more basic than an sp^2 nitrogen, but for guanidine, protonation of the sp^2 nitrogen occurs because this forms the more stable conjugate acid (stabilised by resonance).

The electron donating propyl group (+I) in 1-aminopropane increases the basicity and it stabilises the positively charged nitrogen in the conjugate acid.

Section 1.7.2 shows the resonance forms of aniline

For aniline, the lone pair on nitrogen is delocalised on to the benzene ring, which decreases the basicity (i.e. the benzene ring helps to "tie up" the lone pair making it less basic). On protonation of aniline, the positively charged conjugate acid cannot be stabilised by delocalisation.

The resonance forms of ethanamide are shown in Section 1.7.2

Ethanamide is the weakest base because of delocalisation of the nitrogen lone pair with the electron-withdrawing carbonyl group.
   (b) 4-methoxyaniline > 4-methylaniline > aniline > 4-nitroaniline

Section 1.7.2 discusses the basicity of 4-methoxyaniline, 2-nitroaniline and 3-nitroaniline

For 4-methoxyaniline, the electron donating OMe group (+M) increases the basicity of the aniline NH_2 group more than the 4-methyl group (+I) in 4-methylaniline. Both of the 4-substituted anilines are more basic than aniline as they contain electron-donating groups at the 4-position of the ring. In contrast, the strongly electron withdrawing nitro group (−I, −M) reduces the basicity of the NH_2 group and so 4-nitroaniline is much less basic than aniline.

7.

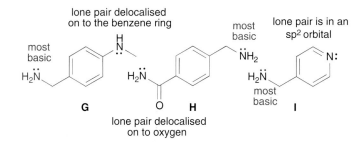

D
H is bonded to an electronegative oxygen; the conjugate base (carboxylate ion) is stabilised by resonance

E
Deprotonation of either H gives a conjugate base that is stabilised by the −I and −M effects of both C=O groups

F
H is bonded to an electronegative oxygen

The resonance forms of a carboxylate ion are shown in Section 1.7.1

8.

lone pair delocalised on to the benzene ring

most basic

lone pair is in an sp^2 orbital

G

lone pair delocalised on to oxygen

H

most basic

I

most basic

Compound **I** contains a pyridine ring (Sections 1.7.5 and 7.10)

9. (a) Product side
(b) Reactant side
(c) Reactant side
(d) Product side
(e) Product side

Chapter 2

1.

(a)

(b)

(c)

Skeletal structures are introduced in Section 2.5

The carbon chains adopt a zigzag structure (Section 3.2.2)

(d)

(e)

(This could be *E* or *Z*, see Section 3.3.1)

(f)

(This could be *E* or *Z*, see Section 3.3.1)

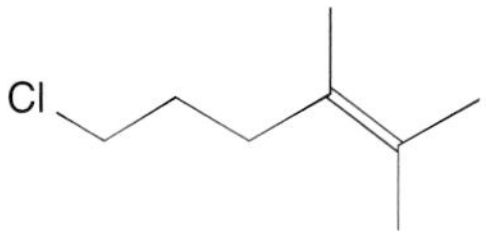

(g)

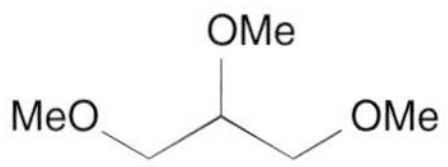

(h)

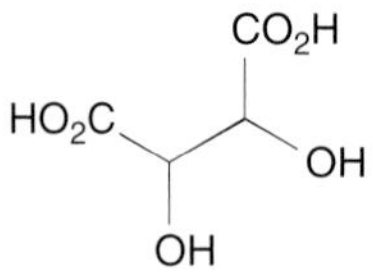

(i)

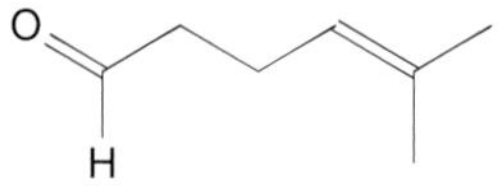

(j)

For naming carbon chains, see Section 2.4

Compound (a) is the product of an aldol reaction (Section 8.5.1)

Compound (c) is an example of a tertiary alcohol (Section 2.1)

Compound (e) is an example of an allylic halide (Sections 5.3.1.2 and 5.3.1.9)

Compound (g) is an example of a secondary amide (Section 2.1)

Compound (i) is an example of a tertiary amine (Section 2.1)

Alkyl substitution is discussed in Section 2.3

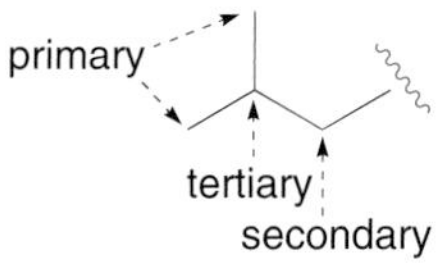

Acid (acyl) chlorides, RCOCl, are introduced in Section 2.1

2. (a) 3-Hydroxy-2-methylpentanal

(b) Ethyl 4-aminobenzoate

(c) 1-Ethynylcyclohexanol

(d) 4-Cyanobutanoic acid

(e) 3-Bromo-1-methylcyclopentene

(f) 5-Methyl-2-(1-methylethyl)phenol or 2-isopropyl-5-methylphenol (isopropyl = 1-methylethyl = $CHMe_2$)

(g) N-1,1-Dimethylethylpropanamide or N-*tert*-butylpropanamide (*tert*-butyl = 1,1-dimethylethyl = CMe_3)

(h) 4-Chloro-6-methylheptan-3-one

(i) Benzylcyclohexylmethylamine (benzyl = phenylmethyl = $PhCH_2$)

(j) 2,5-Dimethylhex-4-en-3-one

3. (a)

primary

tertiary

secondary

(b) Ethanoyl chloride

(c)

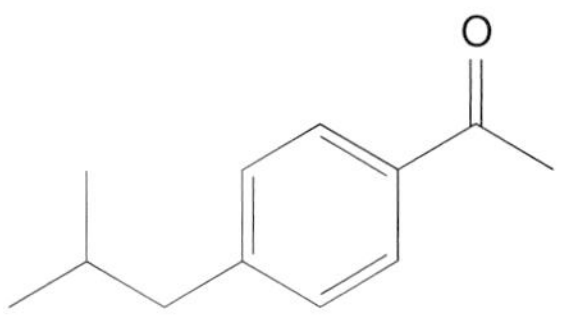

A

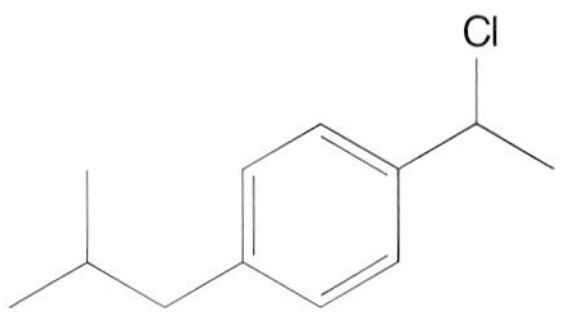

B

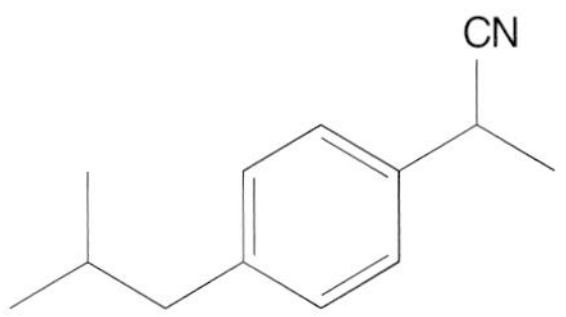

Compounds **A**-**C** are all examples of disubstituted benzenes

C

(d) 2-(4-Isobutylphenyl)propanoic acid

4. (a)

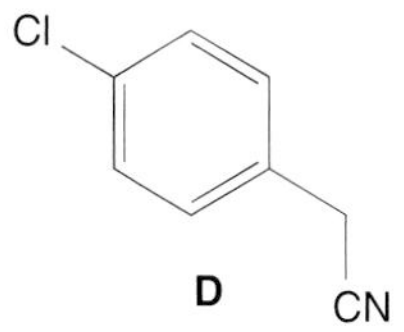

D

Compound **D** is a 1,4-disubstituted benzene

(b) 2-Chloropyridine (**E**)

The basicity of pyridine is discussed in Section 1.7.5

(c)

A tertiary amine

F

Tertiary amines are good bases (Section 1.7.2)

(d)

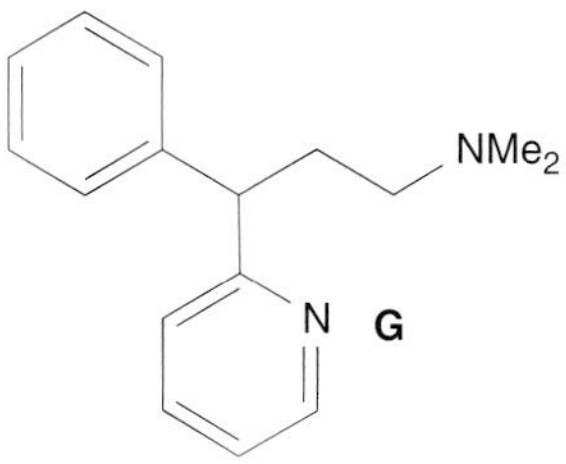

G

Compound **G** has a chiral centre (Section 3.3.2.1)

Chapter 3

An OH group attached to a C=C bond is an enol (Section 8.4.1); an OR group attached to a C=C bond is an enol ether

1. (a) –I (highest), $-CH_3$ (lowest)
   (b) $-CO_2H$
   (c) **A** = (*E*), **B** = (*E*), **C** = (*Z*)

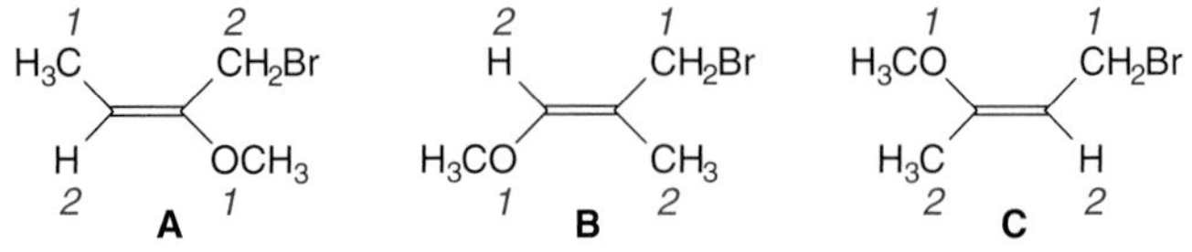

1
H3C
2
CH2Br
H
2
OCH3
1
A
2
H
1
CH2Br
H3CO
1
CH3
2
B
1
H3CO
1
CH2Br
H3C
2
H
2
C

Compound **E** is an amino acid (Section 11.3)

2. **D** = (*S*), **E** = (*R*), **F** = (*R*)

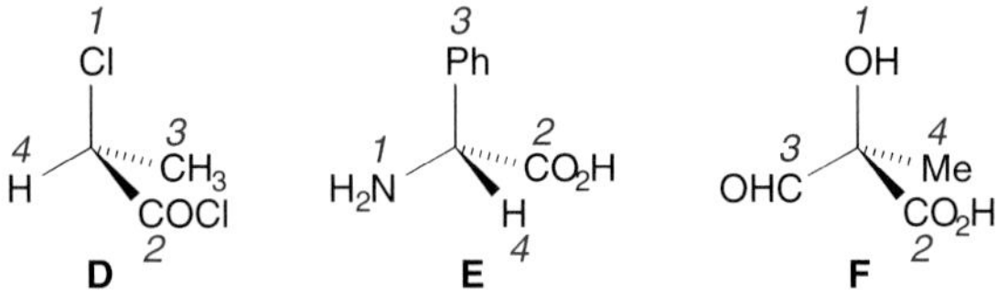

1
Cl
4
H
3
CH3
COCl
2
D
3
Ph
1
H2N
2
CO2H
H
4
E
1
OH
3
OHC
4
Me
CO2H
2
F

3. (a) It is optically inactive but contains stereogenic centres. As **G** has a plane of symmetry, the optical activity of one half of the molecule cancels out the other half.

A compound with two OH groups is called a diol; diols can be formed by dihydroxylation of alkenes (Sections 6.2.2.6 and 6.2.2.7)

   (b)

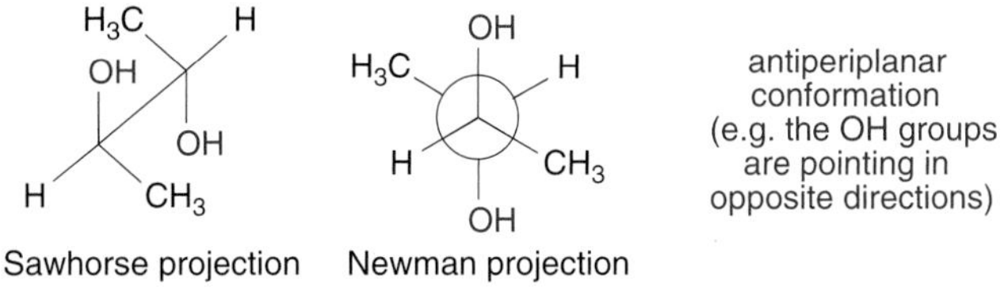

H3C
H
OH
OH
H
CH3
Sawhorse projection
OH
H3C
H
H
CH3
OH
Newman projection
antiperiplanar conformation (e.g. the OH groups are pointing in opposite directions)

   (c)

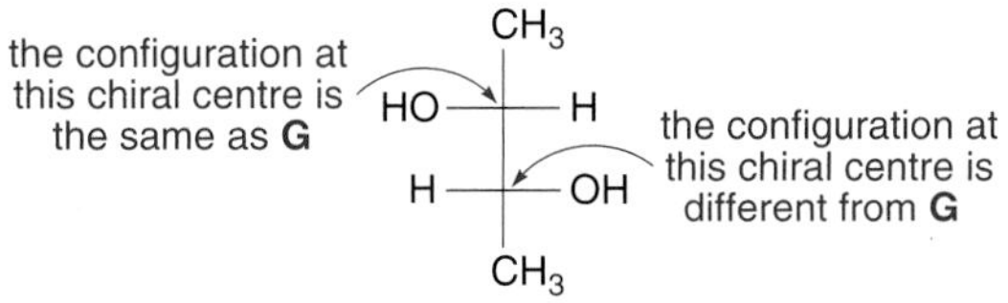

the configuration at this chiral centre is the same as G
CH3
HO
H
H
OH
CH3
the configuration at this chiral centre is different from G

The IUPAC name for menthol is 2-isopropyl-5-methylcyclohexanol (Section 2.4)

4.

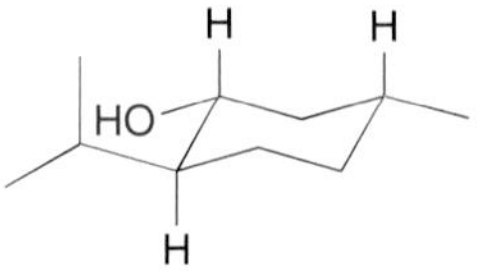

H
H
HO
H

All three substituents are equatorial

5.

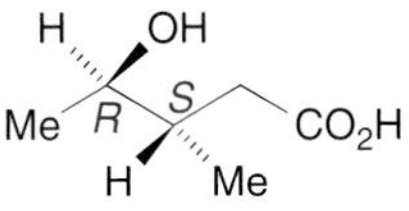

I

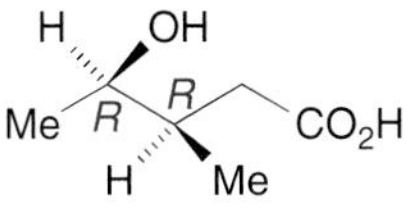

J

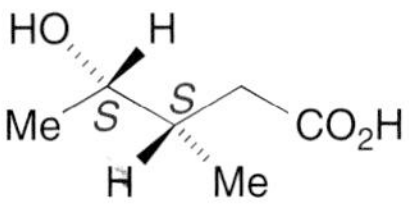

K

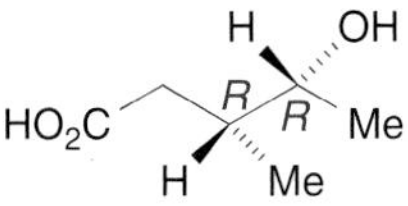

L

These compounds are all 4-hydroxy-3-methylpentanoic acids (Section 2.4)

I and **J** are diastereoisomers
I and **K** are diastereoisomers
I and **L** are diastereoisomers
J and **K** are enantiomers
J and **L** are identical
K and **L** are enantiomers

6.

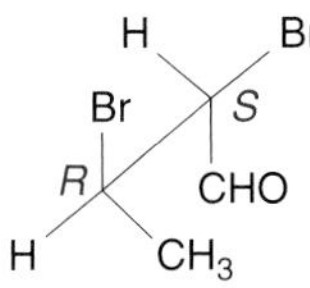

M

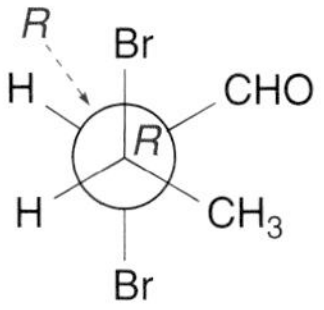

N

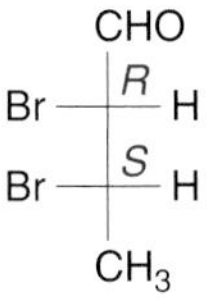

O

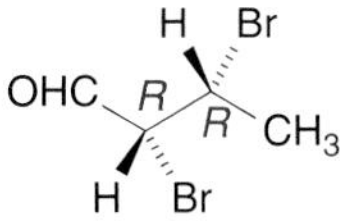

P

1,2-Dibromides or vicinal dibromides are formed on bromination of alkenes (Section 6.2.2.2)

M and **N** are diastereoisomers
M and **O** are enantiomers
M and **P** are diastereoisomers
N and **O** are diastereoisomers
N and **P** are identical
O and **P** are diastereoisomers

7. (a)

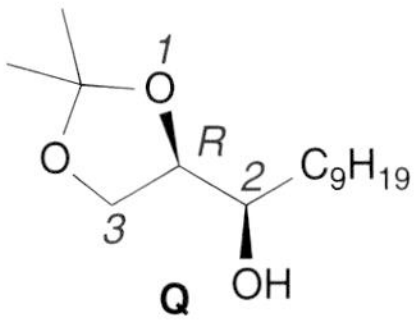

Q

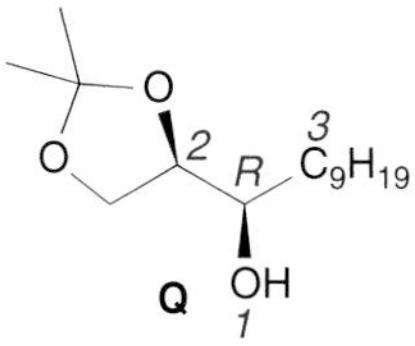

Q

Compound **Q** contains a cyclic acetal; acetals are commonly used as protecting groups in synthesis (Section 8.3.5.2)

(b)

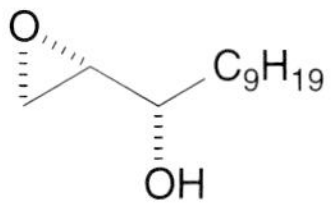

Epoxides can be formed by epoxidation of alkenes (Section 6.2.2.6)

Organolithium reagents (RLi) are strong bases and nucleophiles (Section 8.3.4.2)

(c)

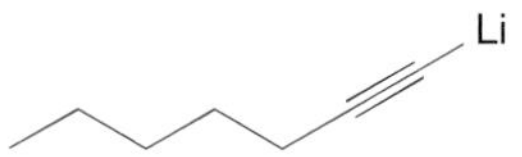

Formation of tosylates (ROTs), from alcohols, is discussed in Section 5.2.2

(d)

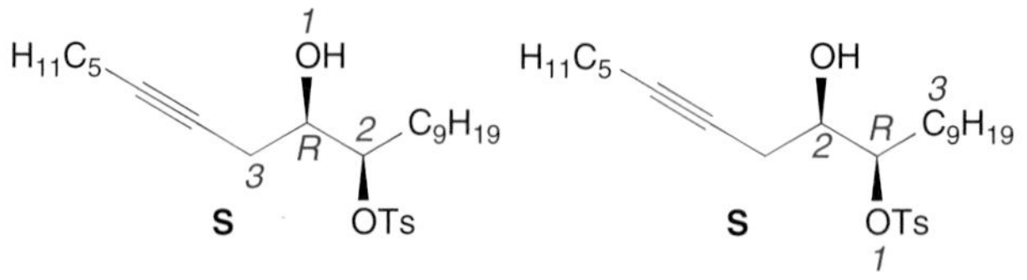

(e)

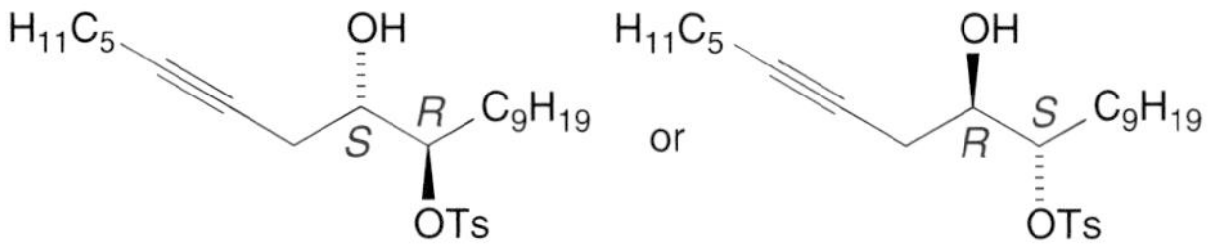

Hydrogenation of alkynes to form (Z)-alkenes is discussed in Section 6.3.2.4

(f)

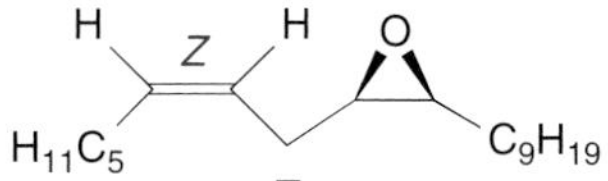

(g)

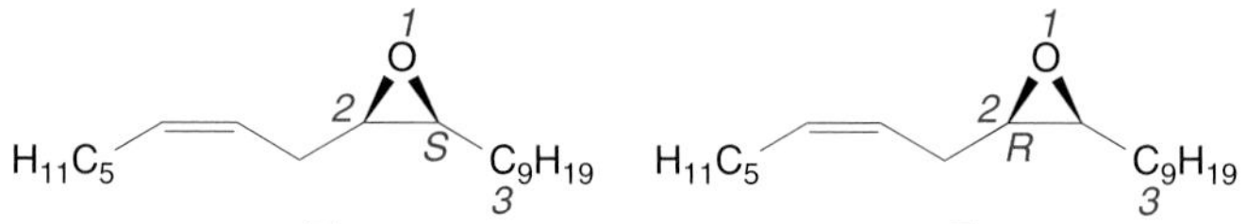

Chapter 4

1. Oxidation levels are shown in italic (see below).

Hydrolysis of 2-bromopropane forms 2-propanol

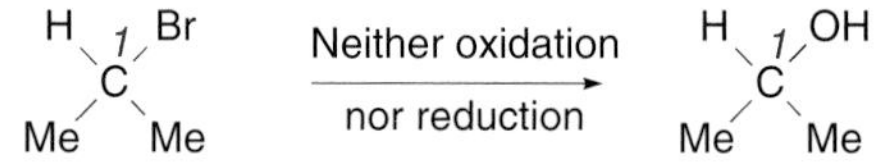

An acetal is converted into a ketone (Section 8.3.5.2)

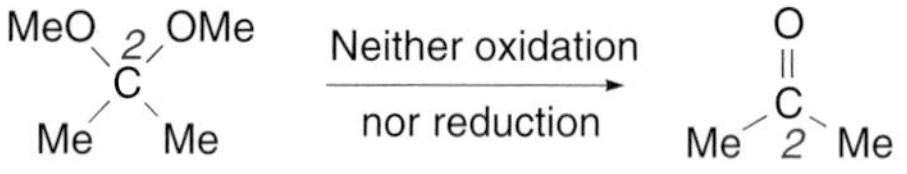

Reduction of an imine (Section 8.3.7.2)

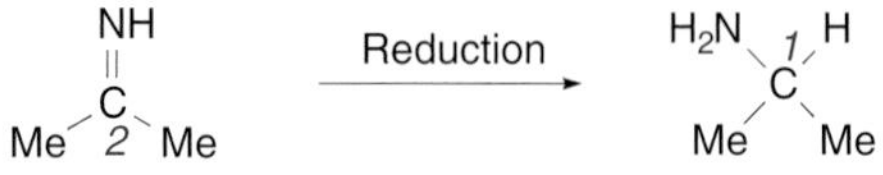

Transformation of an ester into a ketone

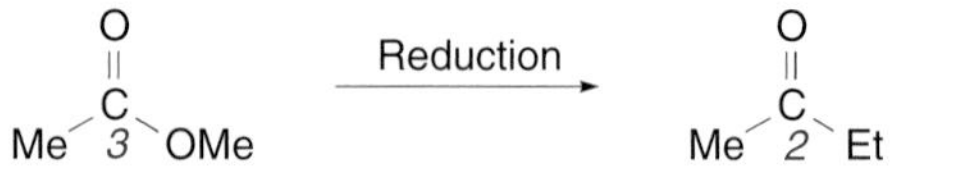

Transformation of a primary alcohol into a secondary amide

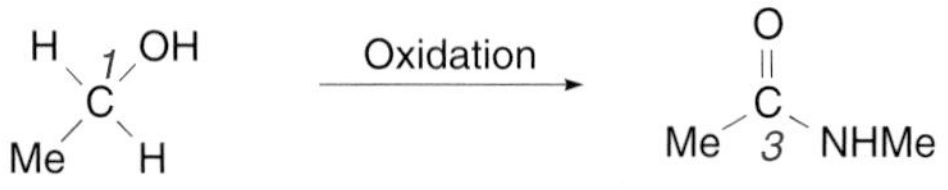

2.

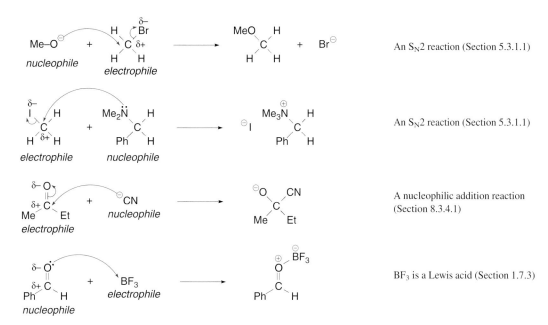

An S_N2 reaction (Section 5.3.1.1)

An S_N2 reaction (Section 5.3.1.1)

A nucleophilic addition reaction (Section 8.3.4.1)

BF_3 is a Lewis acid (Section 1.7.3)

3. (a) (radical) substitution
   (b) addition
   (c) two eliminations
   (d) (Beckmann) rearrangement

Radical halogenation (Section 5.2.1)

Bromination of an alkene (Section 6.2.2.2)

The Beckmann rearrangement is discussed in Section 8.3.7.2

4. (a) stereoselective (forming the *E*-isomer from the elimination)
   (b) regioselective (elimination to give the less substituted alkene)
   (c) chemoselective (reduction of the carboxylic acid)
   (d) stereoselective (reduction of the ketone to selectively form one diastereoisomer)

For eliminations see Section 5.3.2

5. (a)

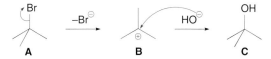

This is an S_N1 reaction (Section 5.3.1.2)

(b) Carbocation **B** (a tertiary carbocation) is more stable than $CH_3CH_2^+$ (a primary carbocation). There are three electron-donating (+I) methyl groups in carbocation **B**, compared to only one electron-donating methyl group in the ethyl cation. Carbocation **B** also has greater steric hindrance – the three methyl groups shield the carbocation (thereby reducing its reactivity) more effectively than one methyl and two (smaller) hydrogen atoms in the ethyl cation.

Tertiary carbocations are more stable than primary carbocations due to electronic and steric effects (Sections 4.3 and 4.4)

The HOMO (lone pair on oxygen) overlaps with the LUMO (empty p orbital on carbon)

(c)

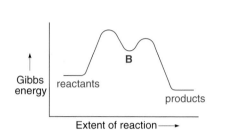

(d)

i.

The two maxima correspond to two transition states, the minima to a reaction intermediate

ii. It is an exothermic reaction (that gives out energy and heats the surroundings) as the products are at a lower energy than the reactants.

For an exothermic reaction, the surroundings get hotter

iii. The conversion of **A** into **B** is the rate-determining step as this step has a larger Gibbs energy of activation ($\Delta^{\ddagger}G$), than for the conversion of **B** into **C**.

6. (a)

This is an electrophilic addition reaction (Section 6.2.2.1)

The HOMO is the π bond of the alkene and the LUMO is the σ* orbital of HCl

(b)

filled π orbital

empty σ* orbital

D

This reaction involves a hydride shift (Section 6.2.2.1)

(c) Compound **E** is a secondary carbocation and this rearranges to form a more stable tertiary carbocation, compound **F**. A tertiary carbocation is stabilised by three electron-donating (+I) alkyl groups whereas a secondary carbocation is stabilised by only two.

The HOMO is the σ orbital of the C–H bond and the LUMO is the empty p orbital of carbocation **E**

(d)

empty p orbital

filled σ orbital

E

Chapter 5

1. (a)

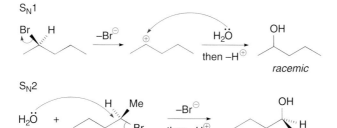

This compound is (*R*)-2-bromobutane

(b) Ethanol can act as a nucleophile and react with the secondary haloalkane to form an ethyl ether (see below) together with HI. As ethanol (EtOH) is a polar protic solvent the reaction is likely to involve an S_N1 reaction to give a racemic product.

A racemic product is a 1:1 mixture of enantiomers

This compound is 2-ethoxybutane

(c) Hydroxide is only a moderate nucleophile but a good base. Elimination can therefore compete with substitution leading to the formation of alkene products.

Substitution versus elimination (Section 5.3.3)

(d) For an S_N1 reaction, the rate-determining step is the formation of a carbocation. Whereas **A** produces a secondary alkyl carbocation, S_N1 reaction of (1-iodoethyl)benzene produces a benzylic carbocation, $PhCH^+CH_3$. This benzylic carbocation is stabilised by resonance and should be formed more readily than $EtCH^+CH_3$.

The phenyl group is a +M group; the methyl group is a +I group

2. (a)

Br has the highest priority, then the propyl group, the methyl group and finally H (Section 3.3.2.2)

(b)

S_N1

Loss of Br^- forms a secondary carbocation

S_N2

An S_N2 reaction leads to inversion of configuration

(*R*)-2-bromopentane   (*S*)-pentan-2-ol

(c) Using a polar non-protic solvent such as dimethyl sulfoxide (CH_3SOCH_3) could promote an S_N2 reaction.

Dimethyl sulfoxide has the abbreviation DMSO

(d) This is an example of nucleophilic catalysis, which exploits the fact that I^- is a good nucleophile and a good leaving group. Reaction of **B** with the iodide ion in an S_N2 reaction produces 2-iodopentane, which reacts more rapidly with water than does 2-bromopentane (**B**).

I^- is a better leaving group than Br^-

(e) Sodium *tert*-butoxide is a poor nucleophile but a strong base and so elimination rather than substitution is favoured. As Me_3C-O^- is a bulky

This is an example of a regioselective elimination

base, pent-1-ene is likely to be the major product (from a Hofmann elimination).

Protonation also forms a stronger (positively charged) electrophile

(f) On protonation, the OH group of **C** is converted into a better leaving group, namely water. Loss of water produces a carbocation, which can rapidly react with ethanol in an S_N1 reaction.

Reaction (a) forms an ether; reaction (b) forms an ester

3. (a) $PhCH_2OMe + HBr$
(b) $PhCH_2OCOCH_3 + HBr$

Reaction of an alcohol with SOCl₂ forms a chloroalkane

4. The mechanism of the reaction changes from S_Ni in the absence of pyridine to S_N2 in the presence of pyridine (which acts as a base).

The H and Cl groups adopt an antiperiplanar arrangement

5.

(1S,2S)-1,2-dichloro-1,2-diphenylethane   (Z)-1-chloro-1,2-diphenylethane

Conversion of the secondary carbocation into a more stable tertiary carbocation is an example of a 1,2-hydride shift. The name indicates that H⁻ (a hydride ion) moves to an adjacent carbon (Section 6.2.2.1)

6. Ethanol is a poor nucleophile and a weak base and so S_N2 and E2 reactions are unlikely. As it is a polar protic solvent, E1 and S_N1 reactions are favoured and so an initial carbocation can be formed on loss of Br^- (see below). This secondary carbocation can undergo rearrangement to form a more stable tertiary carbocation, which can react with ethanol to give **E** and **F**.

7. (a)

The two H atoms between the two C=O bonds are the most acidic. On deprotonation, the resulting enolate ion is stabilised by delocalisation of the negative charge over both C=O bonds (Section 8.4.3)

(b)

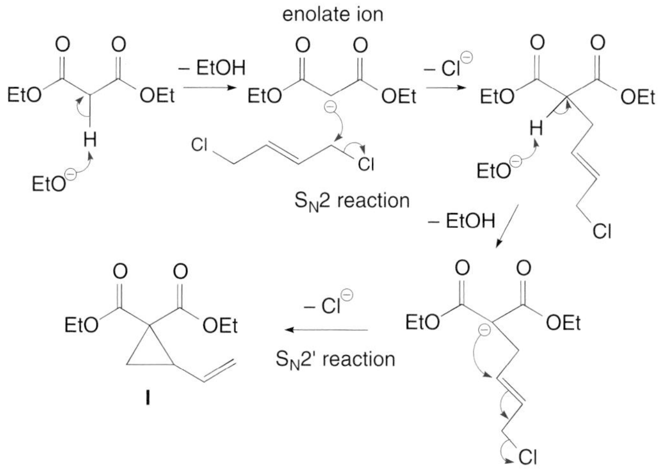

enolate ion

This reaction forms a cyclopropane ring by an S_N2 reaction followed by an S_N2' reaction

S_N2 reaction

S_N2' reaction

I

(c)

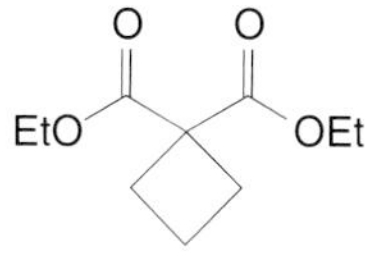

J

This reaction forms a cyclobutane ring by two sequential S_N2 reactions; the first substitution reaction involves loss of Br^- and this is followed by loss of Cl^-

Chapter 6

1. (a) This involves a radical mechanism in which the bromine radical adds to the least hindered end of the alkene to form the more stable carbon-centred radical (i.e. a secondary radical $R_2CH^\bullet$, rather than a primary radical $RCH_2^\bullet$).

Alkyl groups are $+I$ groups that stabilise carbon-centred radicals

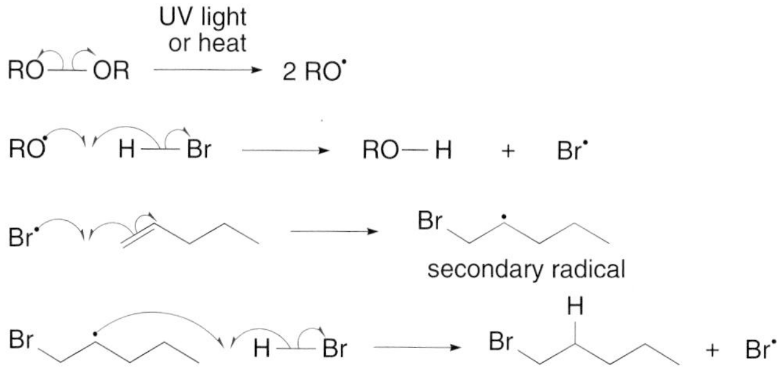

UV light or heat

$RO{-}OR \longrightarrow 2\ RO^\bullet$

$RO^\bullet \quad H{-}Br \longrightarrow RO{-}H \ +\ Br^\bullet$

A stronger O–H bond is formed at the expense of a weaker H–Br bond

$Br^\bullet \longrightarrow$ secondary radical

The product is 1-bromopentane

(b) This involves an ionic mechanism in which protonation of the alkene produces the more stable carbocation intermediate (i.e. a secondary carbocation R_2CH^+, rather than a primary carbocation RCH_2^+).

Alkyl groups are $+I$ groups that stabilise carbocations

The product is 2-bromopentane

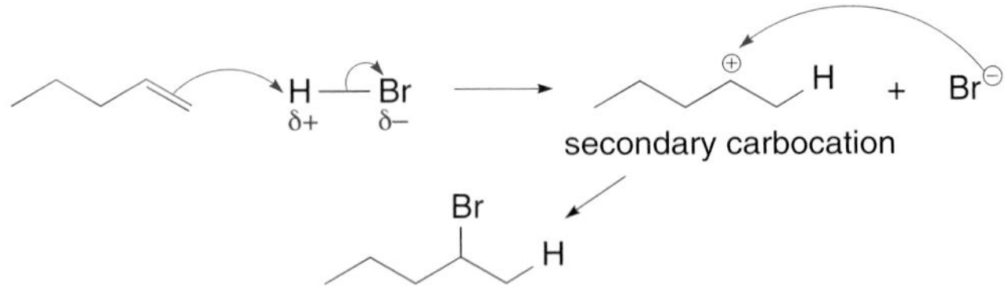

secondary carbocation

Remember that Br$_2$ reacts equally with the top or bottom face of the C=C bond (reaction with the bottom face is shown here)

2. The reaction involves an intermediate bromonium ion, which can be ring-opened by attack of Br$^-$ at either carbon atom of the three-membered ring (as these are equivalent). The bridging bromine atom prevents rotation about the central C–C bond and so the *cis*-relationship of the ethyl groups in the alkene is conserved in the bromonium ion. Overall, there is *anti*-addition of Br$_2$ to give a racemate.

The enantiomers of 3,4-dibromohexane have a different configuration for both chiral centres

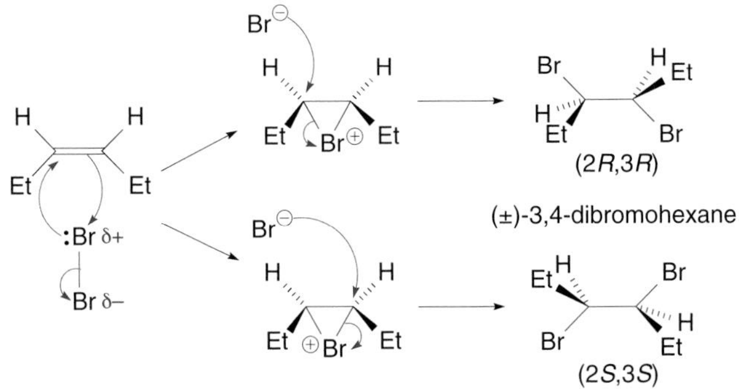

(2R,3R)

(±)-3,4-dibromohexane

(2S,3S)

Epoxides are popular compounds in synthesis; various nucleophiles react with the 3-membered ring to form a range of useful products

3. (a) RCO$_3$H, such as *meta*-chloroperoxybenzoic acid (*m*-CPBA).
   (b) RCO$_3$H and then ring opening of the epoxide using HO$^-$/H$_2$O or H$^+$/H$_2$O to give stereoselective *anti*-dihydroxylation.
   (c) H$_2$/Pd/C.
   (d) OsO$_4$ then H$_2$O/NaHSO$_3$ (to give stereoselective *syn*-dihydroxylation).
   (e) BH$_3$ (regioselective hydroboration) then H$_2$O$_2$/HO$^-$ (H and OH are added in a *syn* manner).
   (f) O$_3$ (ozonolysis) then H$_2$O$_2$ (oxidative workup).

PhCH$_2$Br is called benzyl bromide; PhCH$_2$Br and MeBr are good electrophiles

4. (a) Hg(OAc)$_2$/H$_2$O/H$^+$ (hydration)
   (b) (i) NaNH$_2$ then MeBr; (ii) Na/NH$_3$
   (c) (i) NaNH$_2$ then PhCH$_2$Br; (ii) H$_2$/Lindlar catalyst

5.

The 5-membered ring in part **E** is an example of an aromatic heterocycle called furan (Section 7.11)

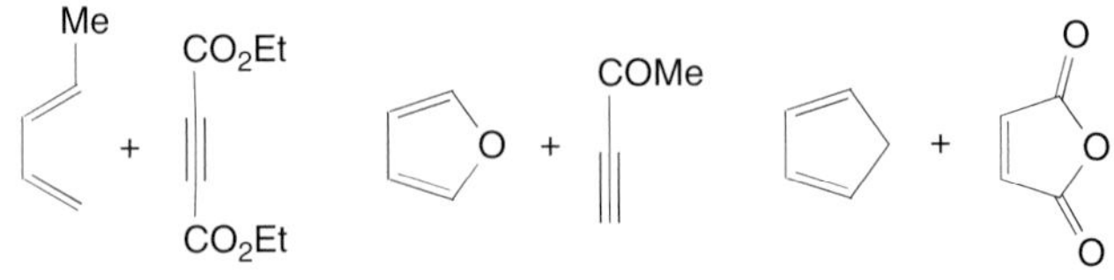

D   **E**   **F**

For compound **F**, the so-called *endo* product is formed. For the *endo* product, the newly formed alkene and carbonyl group(s) are on the same side of the product. *Endo-* products are often formed more rapidly than *exo-*products in Diels-Alder reactions.

exo-product   *endo*-product

These compounds are cyclic anhydrides (they are prepared from maleic anhydride)

6. On protonation of one of the double bonds, an allylic cation is formed and Cl$^-$ can react with this in two possible ways (see below). 3-Chlorobut-1-ene is the expected Markonikov addition product resulting from 1,2-addition. 1-Chlorobut-2-ene is derived from trapping of the cation at the terminal carbon in a 1,4-addition.

An allylic cation is stabilised by resonance (notice the use of the double-headed arrow)

1,2-addition   1,4-addition

The wavy line indicates that the C=C bond has *E*- and/or *Z* configuration

7. (a) Step 1 involves the anti-Markovnikov addition of water to the less hindered C=C bond in **G** (i.e. the bulky borane reacts with the C=C bond in the side chain, not the C=C bond in the ring). The bulky borane adds regioselectively to the least hindered end of the C=C bond, in the side chain, and oxidation using H_2O_2/HO$^-$ produces a primary alcohol. In Step 2, the primary alcohol is oxidised by CrO_3/H$^+$ to form the carboxylic acid group in compound **H**.

The borane is called dicyclohexylborane (Cy$_2$BH)

(b)

carboxylate ion attacks from the 'top' face in an S$_N$2 reaction

iodine reacts with the least hindered 'bottom' face of the C=C bond

This is an example of a stereoselective synthesis; the reaction leads to the selective formation of one diastereoisomer

Chapter 7

1.

Electrophiles include $^+NO_2$ and $^+SO_3H$

2.

Tertiary carbocations are stabilised by three +I groups, whereas primary carbocations are stabilised by only one +I group

primary carbocation

tertiary carbocation

methyl shift

Indole is an example of an aromatic heterocycle

3. (a) **A** (cyclobuta-1,3-diene) is anti-aromatic (4 π-electrons).
    B (anthracene) is aromatic (14 π-electrons).
    C (indole) is aromatic (10 π-electrons).
    D is aromatic (10 π-electrons).

The amino acid tryptophan contains an indole ring (Section 11.3)

  (b) This can be investigated by reaction of **C** with an electrophile. Aromatic compounds undergo electrophilic substitution rather than addition reactions. On reaction with an electrophile such as bromine, indole undergoes an electrophilic substitution reaction to form 3-bromoindole (see below). As for pyrrole and furan, no Lewis acid is required for bromination.

Bromine reacts with indole in the absence of a Lewis acid catalyst

NBS stands for *N*-bromosuccinimide

4. (a) $CHCl_3$ and 3 equivalents of benzene (Friedel-Crafts alkylations).
  (b) (i) $CH_3CH_2CH_2Cl/AlCl_3$ or $CH_3CH_2COCl/AlCl_3$ followed by $Zn/Hg/H^+$;
    (ii) NBS/peroxide (bromination occurs at the benzylic position).
  (c) (i) $Br_2/FeBr_3$; (ii) $CH_3Cl/AlCl_3$ (separate 1-bromo-4-methylbenzene from 1-bromo-2-methylbenzene); (iii) $KMnO_4$ (oxidation of the methyl group).

H_2N-NH_2 is a strong nucleophile called hydrazine (Section 8.3.7)

  (d) (i) $Cl_2/FeCl_3$; (ii) 2 equivalents of HNO_3/H_2SO_4 (to make 1-chloro-2,4-dinitrobenzene); (iii) H_2N-NH_2 (nucleophilic aromatic substitution).

5. (a) The $-NH_2$ group is an activating group and directs the electrophile to the 2-, 4- and 6-positions of the ring. In acid conditions, phenylamine is protonated and the $-NH_3^+$ group is deactivating and directs the electrophile to the 3- and 5-positions of the ring.

The $-NH_3^+$ group is a –I group

(b) The alkene (cyclohexene) is not aromatic and so is a more reactive nucleophile than benzene. In phenol, the −OH group (+M) is a strongly activating substituent and so the ring is more reactive to electrophiles than benzene.

In phenol (C_6H_5OH), the +M effect of the −OH group is stronger than the –I effect

(c) As nitration of phenylamine can occur at the 2-position as well as the 4-position, a mixture of regioisomers is formed. Conversion of the amine group into a larger amide group (by reaction with an acyl chloride) can block the 2-position and increase the proportion of substitution at the 4-position. In the final step, the amide is converted back into the amine (HO^-/H_2O).

The amide group, −NHCOR, is larger than the amine group, −NH_2

6. (a) (i) $NaNO_2$/HCl; (ii) KI.

(b) (i) Br_2 (separate 2-bromophenylamine from the desired 4-bromophenyl-amine; the proportion of 4-bromophenylamine can be increased by the introduction of an amide blocking group, see Question 5c); (ii) $NaNO_2$/HCl; (iii) CuCl (Sandmeyer reaction).

(c) (i) Br_2 (excess) to make 2,4,6-tribromoaniline; (ii) $NaNO_2$/HCl; (iii) H_3PO_2.

The −NH_2 group in PhNH_2 is strongly activating; PhNH_2 reacts with bromine in the absence of a Lewis acid catalyst

7. (a)

The elimination reaction produces an intermediate, highly reactive, benzyne (note that the carbon atoms in the triple bond are not identical – if nucleophilic attack occurs at both carbons then two regioisomers are formed)

(b)

This is an S_NAr reaction. The nitro groups at the 2- and 4-positions of the benzene ring stabilise the intermediate carbanion by resonance

Chapter 8

1. (a)

The aldehyde is reduced to form a primary alcohol.

$NaBH_4$ is sodium borohydride

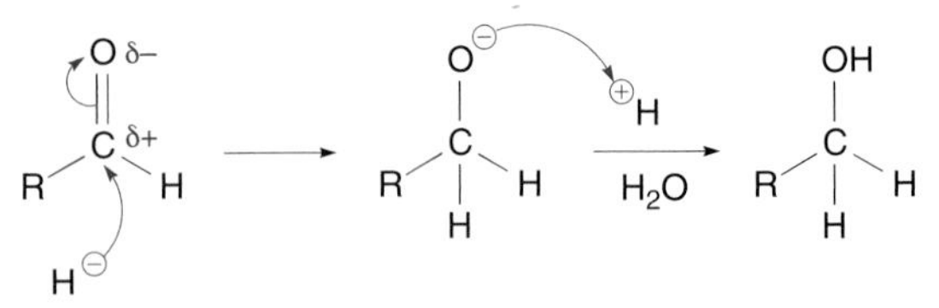

$NaBH_4$ is a source of hydride

PCC has the formula $C_5H_5NH^+$ $ClCrO_3^-$

(b) Use CrO_3/H^+ (Jones oxidation) and distil the aldehyde as soon as it is formed. Alternatively, use a milder oxidising agent such as pyridinium chlorochromate (PCC) to prevent further oxidation of the aldehyde, which would give a carboxylic acid (RCO_2H).

2. (a)

PhMgBr is an example of a Grignard reagent; Ph stands for phenyl, C_6H_5

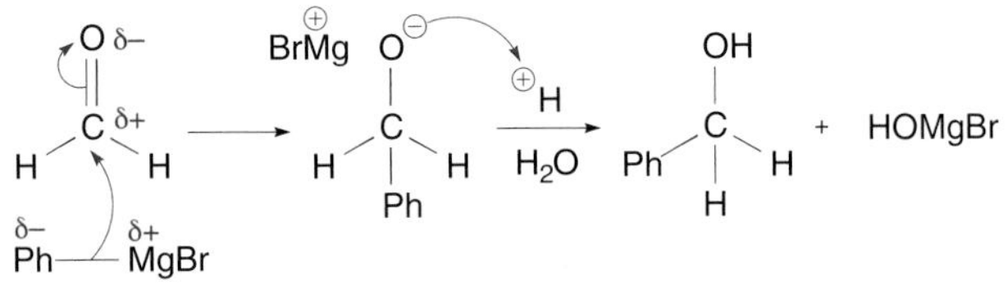

PhMgBr reacts with water to form PhH and HOMgBr

(b) Ph−Br and Mg in anhydrous ethereal solvents (e.g. tetrahydrofuran). Ideally, use an inert atmosphere as Grignard reagents can react with oxygen to form hydroperoxides (ROOH) and alcohols (ROH).

3. $A=CHI_3$ (iodoform) $B=PhCO_2H$ (benzoic acid)

The absorption band at $1700\,cm^{-1}$ is due to the C=O stretch of the carboxylic acid group of benzoic acid

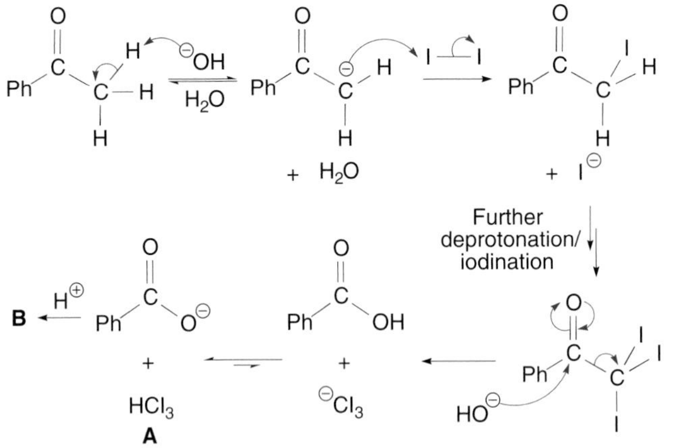

4. (a)

Compound **C** is a cyclic acetal

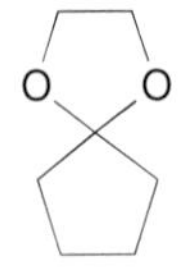

C

(b) The acid protonates the ketone making it a stronger electrophile. This makes it more susceptible to nucleophilic attack by an alcohol.

An electrophile with a positive charge is a stronger electrophile than its conjugate base

(c)

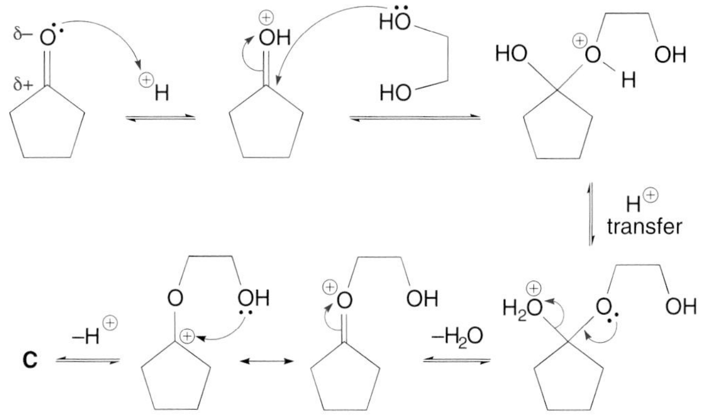

Protonation of the OH group converts it into a good leaving group, namely water

5. (a)

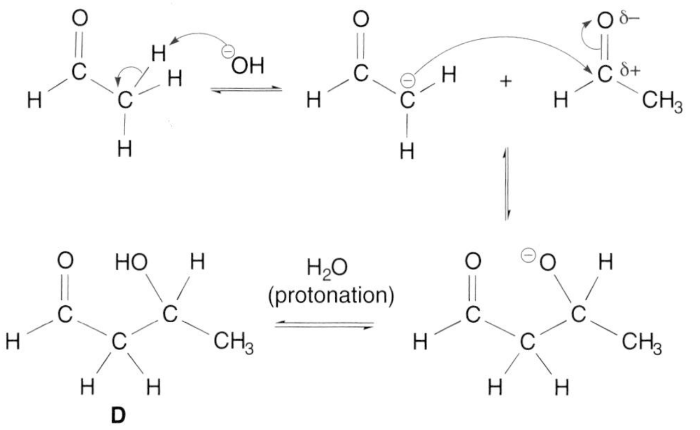

This is an example of an aldol reaction – the product (3-hydroxybutanal) contains both aldehyde and alcohol groups

(b)

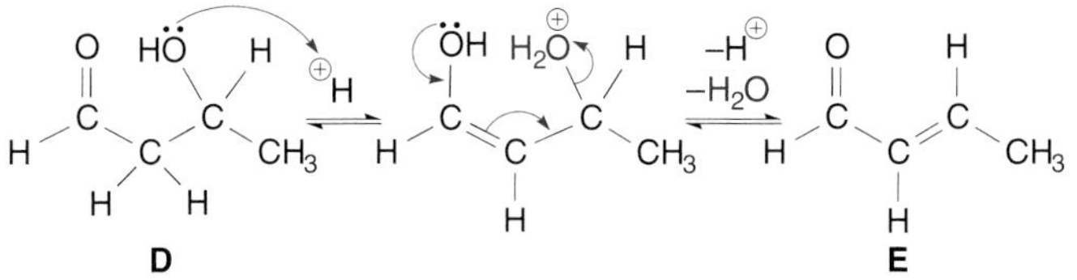

Elimination of water, which involves an intermediate enol, forms the *E*-isomer of an enal (compound **E**)

6.

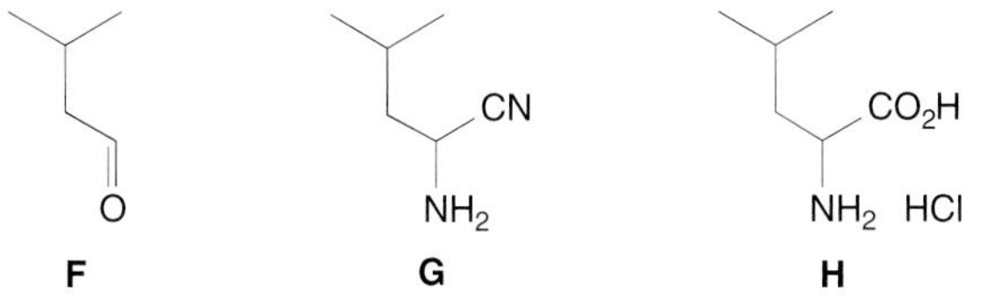

Compound **G** is an amino nitrile and compound **H** is the hydrochloride salt of an amino acid

This aldehyde is commonly called pivaldehyde

7. (a) $(CH_3)_3CCHO$, HO^-

(b)

D   *E*-isomer

The thermodynamically more stable *E*-isomer is formed

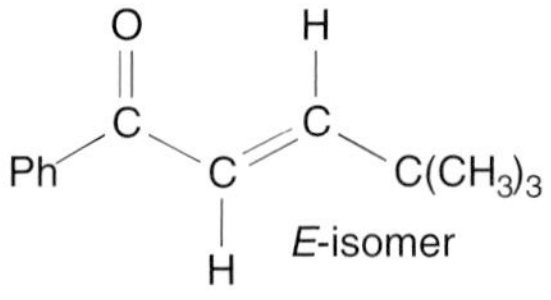

(c) An oxidising agent such as CrO_3/H^+, or PCC.

(d)

Intramolecular hydrogen-bond

A conjugated compound has alternating single and double bonds

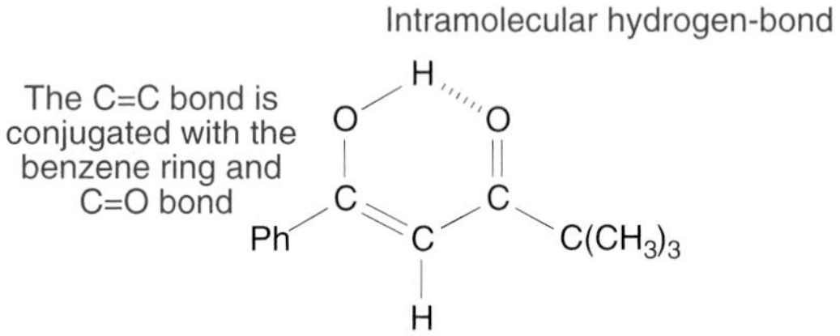

The C=C bond is conjugated with the benzene ring and C=O bond

Chapter 9

1. (a) $NaBH_4$ followed by H^+/H_2O.

The positive mesomeric effect (+M) of the OMe group ensures that the carbonyl carbon atom of the ester is less electrophilic than that of the ketone. Therefore, the ketone is more readily attacked by nucleophiles and mild hydride reducing agents such as $NaBH_4$ only react at the ketone.

The ketone is reduced to form a secondary alcohol using a complex metal hydride

(b)

EtOH acts as a leaving group

This is an example of an intramolecular transesterification reaction catalysed by an acid

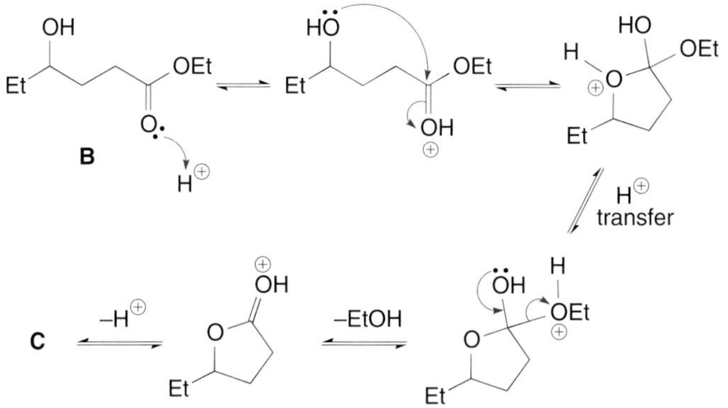

Compound **D** is a diol with one chiral centre

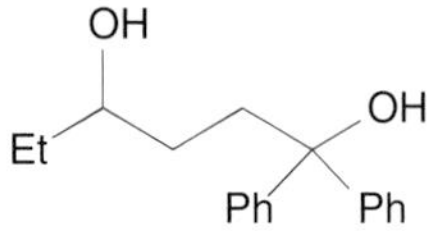

(c)

D

2. (a) H^+/H_2O.

Reduction of the carboxylic acid requires $LiAlH_4$ ($NaBH_4$ is not a powerful enough reducing agent)

(b) $LiAlH_4$ then H_2O.

(c) CrO_3/H^+ (to form $PhCO_2H$) then $EtOH/H^+$ (esterification).

(d) EtOH/H$^+$ (esterification).

(e) H$_2$O.

(f) H$^+$/H$_2$O/heat.

(g) SOCl$_2$ or PCl$_3$.

(h) HNEt$_2$.

HNEt$_2$ is called diethylamine

3. (a) PhCOCH$_2$COPh.

(b) PhCOCH$_2$CHO.

(c) (CH$_3$)$_3$CCOCH$_2$CO$_2$CH$_2$CH$_3$.

(d) PhCOCH$_2$COCH$_3$.

These products are derived from intermolecular Claisen condensation reactions

4.

E   F   G   H

Compound **H** is an example of a hydrazone

5. (a) 1. CrO$_3$/H$^+$. 2. MeOH/H$^+$.

(b) MeO$^-$/MeOH.

(c)

H$_2$C=CHCOCH$_3$ is called methyl vinyl ketone (MVK)

The first formed enolate ion is an example of a soft nucleophile (Section 8.5.4)

This sequence of reactions (i.e. Michael addition followed by intramolecular aldol condensation) to form a 6-membered ring is called the Robinson annelation. Annelation refers to the formation of a ring

Loss of water forms an enone, stabilised by conjugation

Compound **O** is called 1,3-cyclohexanedione

6.

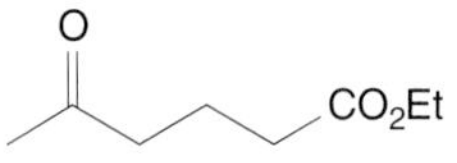

M

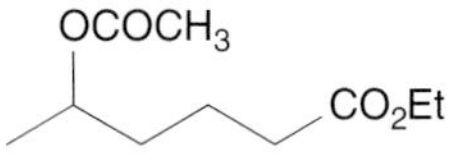

N

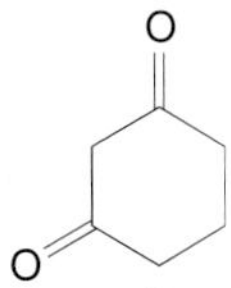

O

Chapter 10

Compound **A** is called 1-(4-bromophenyl)ethanone

1. (a) Use high-resolution EI or CI mass spectrometry.

(b) From the characteristic 1:1 ratio of (M) and (M+2) peaks in the mass spectrum.

(c) A singlet at ~2.6 ppm due to the three equivalent hydrogens of $-COCH_3$ would be observed in the ^1H NMR spectrum.

In the ^{13}C NMR spectrum, a characteristic ketone (–CO–) peak would be observed around 200 ppm (the methyl carbon would give a peak at ~25 ppm).

Fragmentation to form $4\text{-BrC}_4H_6C{\equiv}O^+$ could occur in the mass spectrum. This would help to confirm the presence of an aromatic ketone and the molecular ion peak could be used to establish that **A** is a methyl ketone.

The bromine is at the 4-position, or *para*-position, with respect to the methyl ketone group

(d) Two sets of distorted doublets (with the same *ortho* coupling constant of ~8 Hz) will be observed in the aromatic region of the ^1H NMR spectrum. This is because there is a plane of symmetry in the molecule and so two sets of hydrogens are chemically equivalent. These are shown as H_A and H_M below. The $-COCH_3$ (–M, –I) group is more electron withdrawing than the –Br group (+M, –I) and so the H_M hydrogens have a higher chemical shift. This is known as an AM spectrum because the difference in chemical shift ($\Delta\delta$) of the two doublets is greater than the size of the coupling constant (J). For an AX spectrum, $\Delta\delta$ is much greater than J and for an AB spectrum J is greater than $\Delta\delta$. Notice that as the value of $\Delta\delta$ approaches the value of J so the inner lines increase in height to produce a 'two leaning doublet'.

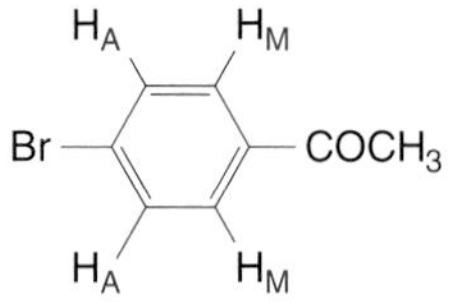

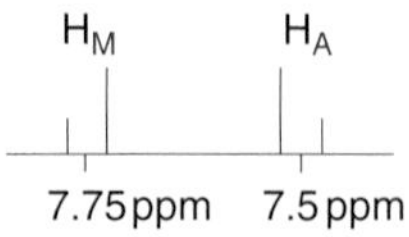

AM spectrum

AX spectrum

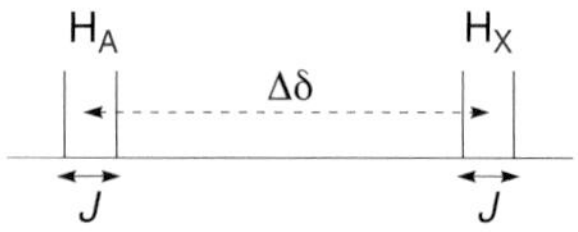

AB spectrum

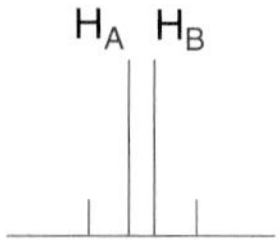

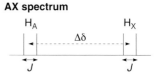

For the 1,2- or 1,3-disubstituted isomer of **A**, there will be four (not two) separate peaks in the aromatic region of the ^1H NMR spectra because none of the aromatic hydrogens are chemically equivalent.

In the ^{13}C NMR spectrum for **A**, there will be four peaks in the aromatic region (two quaternary carbon peaks and two CH peaks). For a 1,2- or 1,3-disubstituted isomer of **A**, there will be six peaks in the aromatic region (two quaternary carbon peaks and four CH peaks) because none of the carbon atoms are chemically equivalent.

2.

Compound **B** is called 1-phenylpropan-2-one

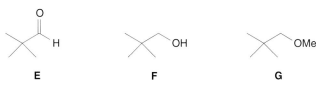

B

3. Approximate chemical shift values for **C** are shown below. Splitting patterns and coupling constants are given in brackets.

^1H NMR

4.1 (q, 7 Hz)

1.8 (d, 7 Hz)

4.5 (q, 7 Hz)   H   Cl

1.3 (t, 7 Hz)

^{13}C NMR

20   55   165 60

Cl

15

Compound **C** is called ethyl 2-chloropropanoate

Approximate chemical shift values for **D** are shown below. Splitting patterns and coupling constants are given in brackets.

^1H NMR

1.8 (t, 7 Hz) 1.0 (t, 7 Hz)

2.1 (s)

CHO

1.3 (sex, 7 Hz)

1.5 (s)   10.0 (s)

^{13}C NMR

210   75   30   20   15

20

10   CHO

205

Compound **D** is called 2-ethanoyl-2-methylpentanal

Compound **E** is called pivalaldehyde

Compound **F** is called 2,2-dimethylpropan-1-ol

4.

E   F   G

Compound **G** is called 1-methoxy-2,2-dimethylpropane

5.

This compound is called pentan-2-one (2-pentanone)

6.

This compound is called 1-ethyl-4-methoxybenzene (4-ethylanisole)

Chapter 11

1.

These dipeptides have the abbreviations Ala-Gly and Gly-Ala

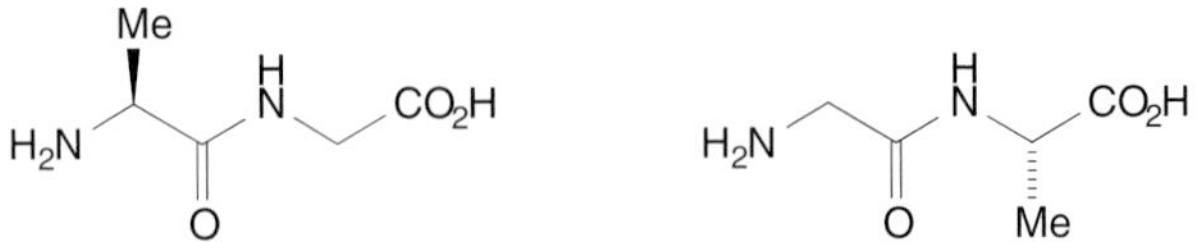

2. (a) Disaccharide **A** contains a β glycoside linkage.
  (b) Two molecules of glucose are formed on hydrolysis.
  (c) Glucose is an aldohexose.

3.

This steroid has 10 chiral centres

Where the rings meet are called ring junctions. If the two groups at a ring junction are on the same side, it has *cis* stereochemistry

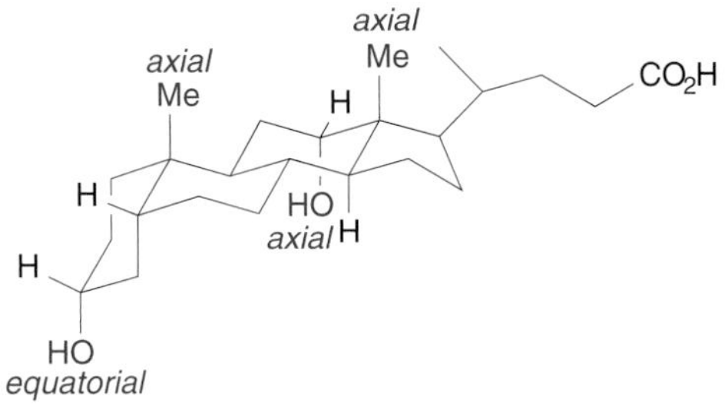

4. This is apparent when tautomeric forms of the amide groups are drawn.

C (uracil)

An aromatic compound is cyclic, planar and has 4n+2 π-electrons

6π electrons

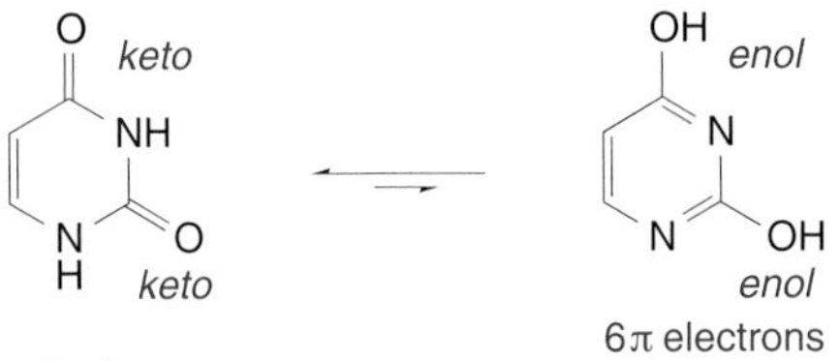

D (guanine)

10π electrons

Addition to the less substituted position is favoured for steric reasons

5. This is because radicals add to the less substituted end (or tail) of the monomer and so the head of vinyl chloride is attached to the tail of another.

PVC has mainly an atactic stereochemistry

head

vinyl chloride

tail

head-to-tail

poly(vinyl chloride) or PVC

6. (a)

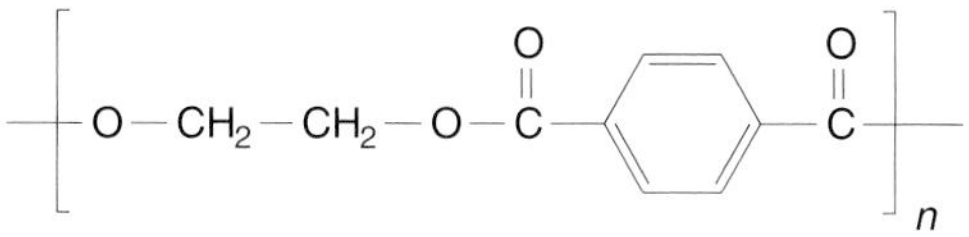

A polyester (Dacron)

This polymer is also called poly(ethylene terephthalate) or PET

(b)

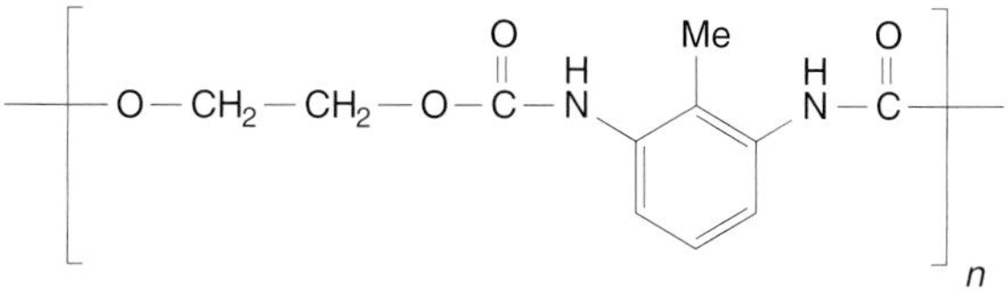

A polyurethane

The C=O bond in a urethane (carbamate) is relatively unreactive to nucleophiles; this is explained by the +M effects of the adjacent O and N atoms

7. (a) Both A and C have hydrogen bond acceptors in the middle position.

(b) In guanine, the nitrogen atoms at positions N-3 and N-7 are the most nucleophilic; the lone pairs on the other nitrogen atoms are all stabilised by resonance, as shown below. Also, in DNA, the 6-membered ring that is part of guanine is involved in base pairing within the double helix and so the NH$_2$ and amide groups are expected to be less accessible to electrophiles.

Notice that all the nitrogen atoms in guanine are sp^2 hybridised

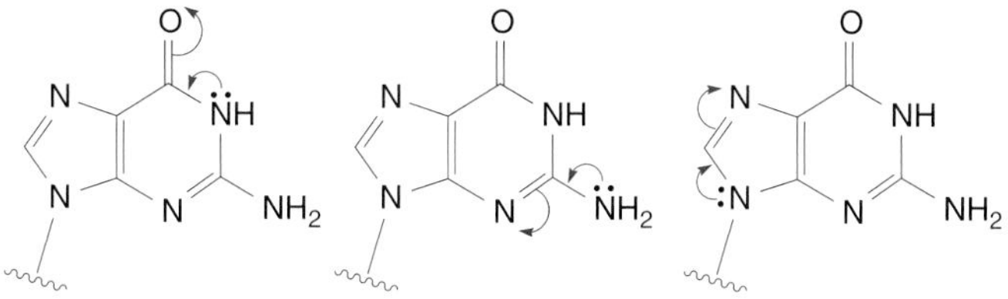

Index

Keynotes in Organic Chemistry, Second Edition. Andrew F. Parsons.
© 2014 John Wiley & Sons, Ltd. Published 2014 by John Wiley & Sons, Ltd.